Troublemakers

The Greatest Dog and Pony Show in Iraq

Stephen Russell

BLUE EAR BOOKS

Published in 2024 by
Blue Ear Books
Seattle, WA
USA

www.blueearbooks.com

ISBN: 978-0-9990951-3-3

Credits:

Cover Design:
Sean Robertson

Book Composition:
Jennifer Haywood, Blue Ear Books, Seattle, WA

CONTENTS

Foreword 1

PART ONE

Chapter 1: The Test 7

Chapter 2: Fort Benning 17

Chapter 3: Basic Training 21

Chapter 4: The Manchu Mile 33

Chapter 5: LSR-D 47

Chapter 6: Dragon Valley 57

PART TWO

Chapter 7: Fort Hood 65

Chapter 8: Step into the Sandbox 81

Chapter 9: King of the Hill 105

Chapter 10: If You're Looking for Trouble 131

Chapter 11: I Saw a Pale Rider 141

Chapter 12: The Wall 151

Chapter 13: Merry Men Are We 157

Chapter 14: Dazed and Confused 165
Chapter 15: Now It's Time to Say Goodbye 173
Chapter 16: Rivendell 181
Chapter 17: Mordor 189
Chapter 18: Death from Above 199
Chapter 19: The Test, Part 2 207
Chapter 20: Don't Rest upon Your Laurels 213
Chapter 21: Back in the Saddle Again 223
Chapter 22: Why Go to Disneyland 233
Chapter 23: Arrowhead Ripper 243
Chapter 24: I Thought I Was Already Home 255
Chapter 25: I'll Never Go Back Home 261

About the Author 267

FOREWORD

For a decade I've wracked my brain over how to present this story. How would those I call my brothers accept or disavow what I had to say? What words might I speak to bring some value of honor, yet tell a very human story that has an impassioned stance on the events in Iraq? There's an almost overbearing feeling of regret when you must add any criticism of events that deal with patriotism and the bond formed by soldiers. I've tried to do my best to remember things correctly and tell the story in a way that doesn't paint people in an unrealistic light. It's been fifteen years since my deployment, and some things are a little fuzzy. There's an idea that you should never betray the brotherhood.

At the same time, I have never been one to toe the line. The more I witness American culture and society decline into chaos and anarchy, the less I care about the grievances or acceptance of my truth. I believe those in positions of authority or power lost touch with reality and the common man long ago. I was born a moody little shit, and I'll die the same. I'm not a hero or an altruistic person you could put on a pedestal or a poster. You wouldn't think I had done anything exceptional, and it shows when I tell anyone snippets about my combat deployment. The events that occurred were crazy and larger than life, and in the media, you aren't going to hear stories from the darkest pits of Iraq. But I feel they must be told. And if not by me, then by whom?

☛

This is the story of a band of brothers like me. We weren't exceptional, we weren't Captain America, we weren't and still aren't heroes. To be fair, your heroes aren't heroes either. Your favorite folk history personages, politicians, scientists, doctors, and artists lied, committed fraud, engaged in treason, were war criminals, owned slaves, had improper relations with children. Jim Bowie was a criminal drunk who illegally sold slaves when slavery was legal. Martin Luther King Jr. had multiple affairs, Thomas Jefferson had sex with his slaves and wouldn't free his mixed children, our country fought a war and killed people to resist taxes imposed on them for starting the French and Indian Wars, which was George Washington's fault by the way. Speaking of George, he sat in a warm room eating and drinking beer as his men nearly froze to death at Valley Forge. Elvis Presley had sex with teenagers, Werner von Braun, who created NASA, was a Nazi scientist and a member of the SS. Your heroes were the villains of progress, savage and disgusting people. This is why I, or anyone else who deployed to Iraq, can't be your hero.

I used to watch pep rallies, sermons, and motivational speakers and wonder why everyone was so enamored. I felt like I was outside the crowd. How is it that people can cling to such bullshit and live with themselves? Where was the brutal honesty of life and society? That life exists in Iraq, and that honesty is why you're alive at the end of the day. It didn't matter if you had heart or were the captain of the football team. It didn't matter if you had a 2.0 GPA or were the class valedictorian. It certainly didn't matter if you were one of Jesus' disciples. Iraq was trying to kill you. You could be the toughest Arnold Schwarzenegger-looking motherfucker or you could look like a limp pile of beans, and it still wouldn't matter. At the end of the day, who was still breathing? Who made it through?

I also don't believe any of us are holier than any others. We deserve what we get, and most of us, to include myself, deserve worse.

We are all so close to becoming the evil we fear. This won't be easy, and I don't plan to make you feel patriotic or ashamed, or any other way military books should make you feel. What I do hope to do is to express the humanity of war. I hope you will come to understand what it's like to be a person experiencing events greater and more intimidating than himself.

I won't pander or adjust fire for your ease of conscience or any other current perception of what you think exists. This will be especially true when it comes to the military and its culture. What you will get is the brutal honesty that was thrust upon me in the Army, and which became an emotional high my very soul began to crave. And I don't pick favorites. This journey is going to require you to step outside your framework and your normal environment. For example, in Iraq many men rape their children. If you think you can make peace with that, then you go over there. On the flip side, if you feel the military might of the United States, brought to you by Haliburton, KBR, Lockheed Martin, and Boeing, is going to instill democracy and equality in the Middle East, you're kidding yourself as well. However, if you think that the people of Iraq are some barbarian hordes of people professing the wrong faith, then I'm not on your side either. The things an Al Qaeda or Taliban terrorist shout are the exact same things I've heard from your good, Christian church on Sunday. Whether you are a Republican or a Democrat, conservative or liberal, society and culture is a thin veneer that we have grown accustomed to.

PART ONE

I HEAR THE CHOPPERS HOVERING
THEY'RE HOVERING OVERHEAD
THEY COME TO GET THE WOUNDED
THEY'RE COMING TO GET THE DEAD

CHAPTER 1

The Test

The sun beat down on me that day in Tahrir like a stepfather's belt on a redheaded child. I had experienced hotter days, where the literal rubber of my boots would melt and leave prints in the cement if I stood too long in one place. I looked behind me as we passed through the gate and counted every IA (Iraqi Army) member as they passed. Twelve men walked in front of me, and each one was my responsibility. SSG (Staff Sergeant) Fili looked at me as I touched each man on the shoulder for my count. He must have thought it asinine to see me imitating what he was doing with the real squad, and I knew why I had the Iraqis in the rear. I was an asshole. I would yell and scream, hit and keep in line any one of them that didn't pay attention or take their role seriously. I would keep them out of the way of real operations. There was a very real threat that any of these men would try to kill us or run away, and it was my responsibility, if those conditions arose, to mow down any of them with my M249/SAW. We knew these illiterate backwoods goat fuckers were from the local neighborhood, and most had been assholes that not six months ago we were fighting in the streets. I'm sure some people at our command outpost (COP) had bounties on their heads, and some of these dudes might feel squirrelly enough to try and collect. We didn't trust any of them, and I would execute them in a second if I smelled a hint of betrayal.

That's why Fili put me where I was, but he made it sound like I was being tested for leadership. The idea that if I could handle some fake soldiers, I might be ready to lead real ones, was what kept me motivated. I didn't know that day would bring out the opportunity I had been waiting almost three years to test. I would finally get to know if everything I had been through from childhood to the first six months of deployment was going to be enough to allow me the fortitude to stand up, kill another human being, and live to do it again. I knew the instant shame I would feel if I couldn't muster the courage. I knew the anger I had held back from birth, the anger of having been born, would be completely illegitimate. I had become the devil I wanted to be. Now it was time to show up and prove it. I was well known by IA groups as an asshole. Some soldiers were disliked by the IA due just to bigotry. They didn't respect them, called them dune coons or sand niggers. I possessed bigotry against them as well, but I felt justified because the average IA soldier was cowardly and unintelligent. You would relay orders through the interpreter, and they would stare at you with those eyes that looked as empty as livestock and not move, or do the opposite of your instructions. I suppose I could have had more empathy, but I was twenty-one years old, we were at war, and these men seemed like they didn't care about fighting or protecting their own country. My reputation specifically came from an incident where I assaulted one of their officers for not paying attention on a patrol. Was that out of line? Of course, but I didn't care. We used to say it's better to be judged by twelve than carried by six. I knew the serious nature of what we were doing, and that fuckface had decided it was more important to put on a show and play Army. He walked out with a uniform that looked more like a British soldier with a pith helmet than someone who was going to fight terrorism. He held his AK behind his back like a riding crop and was literally trying to rock knee-high boots. When he walked forward, he began goosestepping like a Nazi, and at that point I couldn't take it. I began screaming at him until our interpreter, Snake, came to see what was going on.

After I explained what had happened, Snake communicated that the officer was meant to follow me and take orders from me. The officer communicated that he outranked me and would not listen to me. I then explained that if he wouldn't listen and take this seriously, then I would butt-stroke him in the face with my machine gun. I gave this man every chance, at least three times before we were even thirty minutes outside the gate. The fourth time I didn't hesitate, or scream, I just hit the guy with as much force as I could muster and put him on the ground. He screamed and cried and held what looked like a broken and bloody nose, then ran back to the COP without an escort. When we returned from the patrol, our CO (commanding officer) was standing at the gate, looking like an upset mother who was about to tell me to go to my room and wait for my father to come home. But I explained what had happened and the CO understood. After the incident was relayed to the equivalent officer of the IA, he turned around, and that same shitbag was right behind him. The general, as we nicknamed him, began flogging that officer, and we never saw him again. I gained a reputation as being the asshole to the IAs and was always put "in charge" of them, because at least the fear of me kept them from doing anything too stupid.

I tried my best to ensure that each IA soldier held up the standards we were required to hold, and I took my job too seriously. I corrected them on spacing, chastised them for signaling incorrectly, and hit a few any time they flagged the patrol with their rifles. On this patrol SSG Fili and Doc Nguyen were leading our squad. These positions are the most important people in your squad and platoon. Your squad leader is your daddy. He guides you, molds you, gives you life advice, dispenses punishment. Doc, the combat medic, is your maternal figure. He is there not just to care for your wounds but to give you a pick-me-up of an 800mg Ibuprofen when your feet are being a bitch or stick you with an IV after you get too drunk on the weekend. He's also the first guy many people go to when their dick starts burning because they scored with Suzy Rottencrotch over the weekend. SSG Fili was a very tall Samoan who was more obsessed with his looks than R Kelly, even though he had no hair and we all smelled like shit from barely bathing. Doc Ngyuen looked like he could hold his own but was very quiet. He was a new medic from our replacements, and I think he felt he had something to prove.

Next in the squad are usually two team leaders, but we only had one that day. They are usually the big brothers of the group and care for their privates and work with one another to make sure the squad leader doesn't know of all the ill shit you decided to get into over the weekend. At the same time, they beat your ass when you don't fall in line. We were short one team leader that day, but the man available was SGT (Sergeant) Crowley. He was tall and thick with a potato-shaped head. He might have had three good teeth and looked like the love child of Danny DeVito and his extremely tall sister. He had the intelligence of two whole piles of bricks but was the most loyal, honest, and ride-or-die team leader I ever had. SGT Crowley might hate your guts, but he would throw his body on a grenade for you if you were his soldier. The last of the leadership was LT (Lieutenant) PJ. I won't even try to spell his last name. Like many people in the military who have long ethnic names, he ended up getting initials as his name. He was the fifth lieutenant to lead our platoon. Just like the other four, he had something to prove, but at least he seemed calm and well rounded.

The privates that day were Davis, Valverde, and me. Steve the interpreter was also tagging along, but he was new and had replaced another interpreter who had been found leaking information to local terrorists. So far Steve had been relatively helpful, but passive. He had taken a liking to me, and people joked that he was gay for me. Davis and I were best friends and known as the wild boys of the battalion. Our life motto was to be involved in any form of chicanery or ill shit. We drank, cursed, and smoked too much. We were in constant trouble, with multiple Article 15s under our belt, and we demanded to be part of any mission that would involve explosions or that might be a suicidal endeavor. Valverde was the new private from the replacements, a thirty-year-old man trying to gain some lost provisions of youth and add something of meaning to his life. I treated him like a five-year-old child and trained him as hard as I could. At one point he forgot how to take apart an M9 pistol, so I showed him how. After that, when he couldn't remember the steps, I smoked him for ten minutes straight each time he got it wrong. It only took three or four times, but he never forgot after that. He was the second SAW gunner. Steve the interpreter was an Arab from the local town. He enjoyed our company and wanted to bond with us, but we were also cautious.

Our mission that day was to conduct the infamous presence patrol. Later in my Army career I was chastised because "presence patrols" aren't found in the Army doctrine manuals and supposedly don't exist. Yet for some reason every OIF (Operation Iraqi Freedom) veteran remembers the never-ending shifts of these patrols we conducted. Our superiors informed us that the point was to disrupt enemy movement, and that our presence would become more familiar to the local populace. This time felt no different, and we seemed to be making good time through the city of Tahrir. Honestly, I was just trying to beat the heat and finish the patrol. I berated anyone in the formation that looked like they were weakening from the heat. You never could show weakness in that environment. Showing weakness in Iraq was tantamount to death, and you never wanted to look like someone that could be fucked with. Even still, we would get attacked often enough. You would still take fire even though most of the time it was just pop shots, or small bursts meant to disrupt us or test our actions. Sometimes it was meant to draw you into an ambush with IEDs, mortars, or other actions, but rarely would it result in a lengthy firefight between two opposing forces. The enemy was expert at gathering intelligence and would use anything they could to their own advantage. It wasn't long before gunfire began to slam down the street in the middle of our formation, but this was different. This was a purposeful burst of lead. It wasn't a test; someone wanted to kill us. I had been in harm's way before, but never like this. I had been shot at or had poorly aimed bursts of bullets fly over my head. I had not had the rush, the adrenaline, the satisfaction of having them whiz past my head, then choose to fight back. Now my moment had arrived. My ultimate question would finally be answered: Could I do this?

For those of you who have yet to have lead sling past your face, I'll let you in on a little secret: It doesn't sound or feel like you think it would. Everyone sees the explosions in movies and has some misconstrued view about the trajectory of rounds, like they're some magical spell that neglects the laws of physics. In reality, a bullet sounds like a bumblebee. If you hear the crack, or the pop if you will, you are not being shot at; you are behind or to the side of the gun. That's what the movies teach you it sounds like. Yet a bullet traveling through the air sounds like a bee and, in my experience, it angers you like one as well. You want that noise to go away, to stop tickling the sensitive skin on your ear.

The bees began flying past my face and all around me. I jumped to the side wall next to a house and screamed for the IA to about face. They were paralyzed. The rest of the squad was ahead, but it was my job to pull rear security. The last thing you need when you're ambushed is to have another ambush crawl up your ass. I saw the fear spread through the IA soldiers to my front. They didn't know what to do. I ordered them to post on the street behind us and pointed in the direction for them to look. I felt the dirt kick up around me, and the sound of bees grew. I was being targeted. I fell to the prone and rolled to the side.

As I looked on, I saw two Iraqi civilian men on the ground holding the backs of their heads. They were terrified until I started laughing. Hysteria took over and they laughed with me. Yet before I could even take note of anything else, the bees stopped and everyone began looking around. What the fuck just happened? The dust settled like fog on a swamp through thick trees. I could barely see SSG Fili give the signal for everyone to get up. He motioned to me to move the squad of IA behind the street and was already screaming on the radio. I didn't really pay attention, because my job was behind us. Secure the rear was my only thought. I moved through the squad of IA, yelling out positions and arranging them like chess pieces. We all moved to the back alley, then I moved forward to meet SSG Fili. I could hear the rocks crunching beneath my boots, and my sweat drenched my shirt as though I had just come out of the pool with all my clothes on. As Fili and the new LT began arguing over some issue, SGT Crowley ran up to advise them on a new route we could take. I hadn't listened to anything else going on until this moment. Now I needed to know where we were going. I heard them arguing that the building we had taken fire from was covered and concealed on the left-hand side, and that we should use the next alley for cover to move against the asshole that had just ambushed us. That was all I needed to hear. I motioned for the IA to stand up and spread out. Much of the way I communicated with them involved over-emphasizing my facial expressions, while trying to use clear gestures to indicate my commands. The squad moved forward, and I motioned for an IA to move forward and pull security down the street while the squad turned down the next alley. I then turned down the alley and made sure the momentum of the formation was steady, then stayed behind until the last IA soldier crossed in front of me. I made sure no one was following and turned around. I looked up to see the whole squad on either side of the alley, its tall walls closing us in like prison cell blocks. There was wind but no noise. All the gates were closed, and I began to move towards an extended piece of wall. Simultaneously I looked at the building in front of us and realized that it was the same three-story building from the street we had just come from. The GRG (Gridded Reference Graphic, or map) was distorted and made it appear like the building stopped before this alley, but the GRG was wrong.

Before I could shout anything, a burst of rounds began smashing down the center of the road in all directions. The bees returned louder than before. SGT Crowley and Rivera immediately went down. I couldn't see LT PJ or Doc. I could see SSG Fili behind a concrete barrier trying to smash a gate with Davis. No one was firing back! The IA soldiers bolted towards the wall and began trying to hide behind one piece of cover that could barely cover half a person. I pushed one behind me and let my SAW rip, holding down the trigger and aiming at the center of the building. The bees stopped. I let another line of rounds fly and watched dust clouds emerge from the wall. All twelve fucks stood at the side of the wall, pushing each other out in the street in hopes of creating more room behind cover. I screamed at them to start shooting, calling them pussies and telling them to hurry. One man jumped in front of me, and the bees began once more. The next thing I knew there was wetness all over my face. The man in front of me slumped on my chest. His head was open and oozing blood, and I shoved him off my body and spat in disgust. Even through the immediate mortal peril, I could only imagine that I had just contracted AIDS. I threw the man to the ground and continued to shoot. I smiled and prepared to reload.

MAMA, MAMA CAN'T YOU SEE?
WHAT THIS ARMY'S DONE TO ME!
AND IT WON'T BE LONG...
'TIL I, 'TIL I,
'TIL I GET ON BACK HOME

CHAPTER 2

Fort Benning

Arriving at basic training is an interesting experience, to say the least. After going through MEPS (Military Entrance and Processing Station) and allowing an old man to stare at your naked body and check your butthole, you finally get to go through a half-assed ceremony where you raise your right hand and swear by the Constitution to support and defend the American people against all harm both foreign and domestic. The domestic part still gets me, because you can literally observe acts of Congress destroying the rights of fellow Americans and causing all sorts of undue harm, death, destruction, lies, and evil, and the military hasn't blinked an eye. While most Americans were upset by the actions of a few idiots on January 6, 2021, we forget to acknowledge that we force our military members to swear an oath that they will do worse than that if they feel the actions of our government are in direct contradiction to our "holy" constitution. The first official step you take after signing your life away for eight years lacks all supervision or oversight. The same guy who told you to let an old man stare at your butthole gives you a plane ticket and tells you to find someone at the airport to take you to a military base. If you did run away, then you wouldn't even be considered absent without leave (AWOL), as you had never started your contractual obligation to begin with. It is entirely on you to ensure that you make it to the airport and link up with the sergeant at the terminal.

When I arrived in Atlanta to start my training at Fort Benning, I saw two people in uniform and a large group surrounding them. They were corralling people off the plane to gather up in groups to leave on a bus. I remember thinking this was a focal point. I could take the plunge and keep going, or I could run off and start an adventure. I could bolt and never look back. Sometimes I wish I had walked away at that moment. I wish I had been brave and followed my passions to live a life as a transient, bohemian artist. But I didn't. I walked forward and reported for duty.

Once you get there, it's a different story. If you're infantry your first base is Fort Benning. This sad sack of desolate land nestled between Alabama and Georgia has broken many a man over a century and will break many more. Ranger Battalions are located on Benning, Airborne is on Benning, 3rd Infantry Division is there, the Infantry Museum is there. This is infantry land. It was named after some shithead Confederate general, Henry L. Benning, whose ghost still ensures that it rains every time the infantry move out to train in the field. The Infantry Museum proudly displays a painting of the military negotiating for the land with local Indian tribes. The annotation under the painting tells how the Indigenous people gave away the land for next to nothing because it was a worthless swamp infested with horrible snakes and insects. They thought they were getting a deal just getting rid of the land, let alone trading it for anything of value. The infantry looked at this cursed and infested swamp and said, "Here is home."

Upon arrival you're placed in lockdown for inception, monitored every hour of the day, and your entire day is scheduled full of activities like waiting in line to get shots, getting dental work completed, and getting new glasses that cover three-fourths of your face. We called them BCGs, birth control goggles. You chant creeds and information relevant to your new life before eating, sleeping, or doing any activity. Your first meal at Benning is the same meal they've been giving out for thirty years or more. You get a plastic-wrapped Styrofoam plate with a small can of Pringles, a sandwich, a juice cup, and some napkins. Years later, upon my return to Benning, I would be greeted by the exact same first meal and couldn't help but laugh. You're not allowed to move anywhere without another human being, your "battle buddy," being next to you. The showers are completely open, and the stalls are missing from the shitters. Your entire life is exposed for all to see. Some people cannot take this and produce inventive ways to get out of this hellscape. The drill sergeants tell you from day one that the easiest way out of the Army is to finish your time and leave, but some people just do not want to listen. You happen upon people who come forth about lying in MEPS about medical issues, or find some reason to show they "can't" adjust. Some had been there for half the time they were meant to be enlisted in the Army. If they had just done as they were told, they would be halfway through their enlistment and getting ready to exit with full benefits. The only way to leave quickly was to check out on life.

Many trainees tried to run away, and most were caught by MPs within the hour, arrested and processed, then sent back to the transition unit to be legally tried and dishonorably discharged. One guy above my floor couldn't take it anymore and, while I was reading the only book I had (the Army Blue Book), I heard a scream above me, and then heard a thud below me. I looked down to see a young man had nosedived into the common area below. He crushed his head and died instantly. The drill sergeant came out in his brown round and underwear. He began furiously kicking and screaming at the dead kid. He wasn't concerned with this person, or with what had just happened. He was cursing him for making his own life more difficult by causing an investigation and more paperwork.

MY HOME IS FORT BENNING

THE PLACE THAT GOD FORGOT

AND OH, IN THE SUMMERTIME...

THE SUN IS BLAZING HOT!

CHAPTER 3

Basic Training

The rain began to fall on my face in a sheet that quickly became so thick you couldn't see through it. It was hotter than most days, yet I could see my breath as though it were a cold morning. I felt the sweat drip down my face, and I attempted to will it into nonexistence because you dare not move unless you're told to do so. In some sort of innocent faith, I felt I would not be fucked with if nothing could faze me, and the last three weeks had been the worst of my life. And yet the worst was still to come.

I had signed a contract for eight years, just for this moment. A lifetime of anger and hatred was ready to spill forth. What awful, horrible plans I had in store for people across the world that I had never met. Years of abuse both physical and mental were spilling out, ready to be expressed in violence. I lied aloud to everyone that it was about patriotism, vengeance for those who had died on September 11, and any other heroic platitude I could think of. None of this was true. I had vengeance and hate in my heart just because I had been born. As I stood there in the Georgia heat and rain, it was time for me to be educated in pain and anger.

After basic training there were some faceless people about to be killed, abused, and violated because a little boy in Dallas had not been loved by his parents. Because after those parents had spent years expressing their love of drugs and violence, he had had to be adopted by fundamentalist parents who stoked his violent feelings through religious dogma and lies.

The process began with taking everything from me, including my name. Right now I was not a person; I was a number, and my number was 444. I'll never forget that black stenciled number on my OD Green Kevlar helmet that was far too big for my small skull. For the next three months, that would be my name. I would have to earn the right just to be called private, a rank.

I had been standing for what felt like all day and was tired, but I told myself just a little longer. Others had passed out from the heat and fallen out of formation. They would be taken back to the holdover building and wait for the next unit to pick them up. I just kept thinking: A little longer, and I'll get to sit on the bus. That's how you do it: little spurts. They cannot make this last forever. The adventure was now just ten minutes away. We were told we were about to relocate to our new home in Hell, called Sand Hill.

The next thing I noticed was a group of men walking towards the formation. I don't remember faces, but I do remember the news they delivered. There were no buses for our group. The other classes had received buses for transport to their basic combat training (BCT) units but, at just under two miles, we were near enough to walk. I had two duffel bags stuffed to the brim, and a brand-new pair of boots on my feet. I was about to receive the first lesson in the Army: Always expect the worst thing to happen and, if something better happens, you'll be surprised. I had been standing in the heat for hours, and now an entire group of random young men and myself were expected to hoist everything and walk two miles into Hell. The Georgia sun and humidity gave no reprieve as we trudged forward. We were hearing the screams of drill sergeants telling us to stay in line, hurry up, do not fall out! I thought: I've run cross country events that were longer than two miles. I used to ride my bike three or four miles to my part-time jobs. But I had never walked two miles in the Georgia heat, carrying a hundred pounds of gear and personal items. I began shifting my bags and gear to try to take away some of the pain. I could feel the stiff canvas straps of the duffel digging into my arms. I could feel the sweat raining from my forehead. My glasses fogged up, and I began to breath heavily out of my mouth. I could feel the heat rising from the ground, and it felt like the inside of my Kevlar helmet was cooking my head. The leather strap was scraping my chin, and I could feel my socks slipping inside my boots. The blisters that began forming that day would swell, pop, heal, and reform at least a hundred more times before basic training was over.

Looking up, I could see an endless line of trees on either side of me, and the road ahead rose into hills that looked like stairs. It's difficult to judge distance in this land, because you'll march up one hill and then see the hill behind it, looking closer than it really is. This endless mirage is enough to make some go mad. It's not, however, enough to cause anyone to fall out of formation. For if you are the unlucky soul who lets slip his fortitude and endurance, you'll be surrounded by a gang of drill sergeants who will make you regret every life decision you've ever made. Luckily this time, we walked up one hill then turned down the road. I could then see a group of buildings. Everything felt old, and there was a staleness to the air that I had not experienced before. But before I could take in what I was looking at, I noticed a commotion at the front of the formation. We were being funneled, like cattle. Two groups of drill sergeants were screaming at all of us who went through the middle. If you didn't run through, they tripped you and threw you to the back of the line. I witnessed another guy go down, and I'm pretty sure he took a few boots to the face. I just kept running. Eventually the end of the line had a tall drill sergeant who would tell you left or right. Once you ran in the direction he screamed for, there was another who told you to stand in a giant square that had been strategically placed on the ground. Another person ran left instead of right, and he took an arm to the throat and was thrashed to the ground. A few drills picked him up and dragged him to the back of the line to try again.

The indoctrination into violence had begun. From here on out, there was no task that could not be completed in an inherently violent manner. Everything had to have a hundred percent of your effort, motivation, and violence. The time in basic is brief when compared to your contracted time in the Army, but it is the most significant portion of some people's lives. It's easy to see why *Full Metal Jacket* spends the first forty-five minutes just on basic training. Many people experience a rebirth of sorts, a memory of hate, fear, love, anger, and nostalgia. Infantrymen hate Fort Benning, and they love that they came from there. It's something that doesn't leave you, something you spend your life trying to fight against and fix. During the experience you have no time to stop and evaluate what's going on or how it's affecting you. In retrospect, I noticed my description comes from a place of privileged youth. I could feel the clothes and tell they were subpar; the food was disgusting; the buildings and aesthetics were low grade; materials and labor were cheap. My gear was more used than you would find in any military surplus store and extremely uncomfortable, yet I would learn to value it more than gold or a person. This is where you find out your bunkmate has never had a home-cooked meal before, that many guys your age have never worn a suit. The cheap plastic shoes are the only dress shoes some have ever seen. My perspective on this world was about to get shattered in a way I never thought possible. We were all going to be turned into objects, shit, maggots, that did not deserve to breathe the air we were stealing. For some reason, I would spend a year fighting the indoctrination I received during this time. Some misguided feeling of fighting the power, or standing up to authority, or just being a stupid teenager with no life experience caused me to feel I had something in me that could change the world. I had misunderstood my situation, and it took me at least a year to get it through my thick skull that you had to really believe what you were saying or doing. You had to convince yourself, or you wouldn't last.

We were divvied up like poker chips among drill sergeants, and they began taunting us. The drill sergeants laughed at an older man for having a degree, but not using it to become an officer. He was mocked and asked if he had a bullshit degree in underwater firefighting or basket weaving. The drill sergeants pointed out people who looked weird, smoked guys just because of the town or city they were from, took care at every possible moment to remind us what pieces of shit we really were. We were cut off from civilization. We could make one phone call the following Sunday to tell our loved ones that we had arrived at basic training, what our address was, and the date of graduation. If we deviated from a script card given to us, the drill sergeant would hang up the phone and tell you that you couldn't make another call. It was your one chance to speak, and even though my parents begged for some sliver of information or small talk, I only read the script and hung up the phone.

We had three drill sergeants in basic training. Drill sergeant Paul was a black man with a gold tooth and a swagger in his gait from being shot in the ankle. He Crip walked everywhere but had made his mark in the invasion of Iraq with 2nd Ranger Battalion. He was the man to be feared, especially if you were white. If I could, though, I would thank him for what he taught me. He showed me how to work, and what it's like to continue working even when you know you're going to fail. I learned with drill sergeant Paul how to mentally push through a no-win scenario. He hated me most of all, and I hated and feared him.

Drill Sergeant Fleck, a white guy who loved "Combatives," walked hunched like he was ready to engage in a quick draw duel at any moment. He was loud and full of energy, told dirty jokes, sang dirty cadences, seemed like he was on a 24/7 cocaine high.

Drill Sergeant Burke was a large black man who had played Navy football and never spoke a single word. You might, for instance, finish a training lane from a range only to see Drill Sergeant Burke standing in your way. You would plead for some sort of direction or command, and he would silently stare back at you. You would move out, and then they would wait until enough of you gathered around and might look happy for a moment. This would result in what we termed a gaggle fuck of privates, and the trap would be sprung on us. They would yell and ask why we were gathered and joking around. We would tell them Drill Sergeant Burke told us to go over here! They would reply by smoking the dog shit out of us for lying, because Drill Sergeant Burke never spoke to anyone, including his own mother.

The remainder of basic was the average stories of dehumanization, replayed in the psyche as nostalgia, like a broken record in your brain. You were not allowed to do anything of your own free will. You were fucked up for doing what you weren't told and for doing what you were told. You couldn't put your hands in your pockets, you couldn't look away when you were being yelled at or debased, you couldn't even breathe the wrong way. Any personal space you were used to or needed was stolen from you. You were never allowed to be alone. We were forced to strip naked and walk through the showers like cattle and shit in toilets across from one another with no doors. One night I woke to use the john and decided I would masturbate. I realized it had been a month since I had touched my dick or sat by myself for more than five minutes. Yet when I sat down a single soul came in from the darkness and sat right across from me in the john to take his shit. There were ten other shitters open. He could have chosen any of them, but he sat right across from me. It was some Pavlovian response to the shared trauma we were enduring.

I don't hold any ill will toward these memories or people. This is what's required to prepare people for war. I wouldn't understand just how easy this was until after my deployment, but the shame, guilt, or insecurities you brought here with you were slowly deconstructed so new habits could form you into a soldier.

I don't really remember my basic training battle buddies' names. I don't remember the exact unit off the top of my head, or any other exact details. I can see faces, and I remember feelings, but it feels like it all occurred in a past eon. I remember being in trouble, a lot. I remember getting pneumonia. I remember being a shitty shot, and weak. I had spent my entire time in high school in the theater and music arts. I was not an athlete, and I was not an avid outdoorsman. In basic training my daily emotions swung wildly back and forth, and much of what occurred daily seemed in that moment to be the crux of everything, but was then just as fleeting. I remember completing events like the team tower and the obstacle course. I remember sitting down to eat my food and having only three minutes to finish. We fired our M16s from cement foxholes and sweated every day. Drill Sergeant Paul would read us the obituaries on Sundays, just to identify privates who had died that came from his basic training classes. I ran and marched what seemed like every hill in Fort Benning, and all in vain. Every time I accomplished something, it was never enough. There was never positive reinforcement from the drill sergeants of any good behavior. There was nothing to build you back up. And even still there were funny moments, like when a guy grabbed a drill sergeant's hat and impersonated him, or when we tied a sleeping man under his bunk and the impersonator screamed "At ease," but the other man could not get out from under the bed. Even in Hell you must laugh, because at the end of the day it would have little effect if you weren't allowed to hope for something better.

The assault course is competed on the last week of basic training. You're required to maneuver multiple obstacles while mortars and machine gun fire are created. It's mostly safe if you follow instructions. There's always a story or two of one private who gets too scared and tries to stand up and is shot by gunfire. You just keep moving, you don't stand, you don't walk, you just low crawl. It sucks, and you get dirty and are tired, but everyone keeps moving. There are markers for areas you can't crawl over, and you can hear orders coming from bullhorns and speakers to keep moving. But that doesn't mean the whole course is totally safe. I found myself at one point crawling in an area I wasn't meant to and didn't see the marker, because my NVGs (night vision goggles) were covered in mud. Right before I began to move forward, I noticed a flare go up, and in front of my face was a tube and a flare came flying out sparking ash and cinders all over me. I rolled away just in time to not take the explosion point blank and continued crawling. I was in total fear and could not push myself fast enough. I would be the last to finish the race. You always know you don't have to be first, but never be last.

If there was supposed to be a reprieve to the misery, it wouldn't be found the next day at Sand Hill. The reward you received for passing basic training was the goddamned privilege of waking up in the Army. I would come across other people in my service time who had different jobs, and they would repeat the lie that we all go through the same basic, we're all infantrymen first, or even claim their experiences were harder than mine. They were POGs, pronounced "Pogue," and it stood for personnel other than grunt, a moniker of shame for those unfit for the brotherhood. We did not go through the same training; we did not endure the same suffering. I don't care if your MOS (military occupational skill) had the word combat in it; you didn't march the Three Kings or learn why the grass grows green and why God loves the Infantry. You did not conquer Sand Hill. Therefore, every grunt sleeping in the mud and rain embraces the suck. You pull your woobie close and in your cold, wet, delirious state you remind yourself: "At least I'm not a fucking POG."

Again, it was another day. We woke up at the same ass crack of dawn time for PT, ate chow, and formed up at 0900. We were told to about face, and then about face one more time. It was at this moment we were welcomed to infantry training. We continued to drill and train until dark. Now, Drill Sergeant Paul began excitedly telling us each day how many days we had until the end of the cycle. Those were the days left, he would remark, until we were shipped to Iraq to fucking die. Finally, I made it through to the end of this insane Hell. I felt so much hope each day that this would all come to an end, and we might join the normalcy of life. Once you're in a unit, they'd treat you like a person, and it's just a job, right?

The last week of our training consisted of a two-week-long field training where we were supposed to put into play all the "training" we had received throughout our first weeks in the Army. I say "training" jokingly, because everything you learn in basic is thrown out as soon as you leave. You maintain many of the cultural aspects, courtesies and customs like saluting or standing at attention, but any "tactical" knowledge is thrown out for unit SOPs (standard operating procedures) or experience from those who have been forward. The point of basic is not to teach you how to be a soldier, but to teach you how to learn to be a soldier.

After a few days, I grew sluggish and unmotivated. We had played raids, were sleeping outside with no shelters, and had a few people sent to the rear for brown recluse bites, but nothing very intense had occurred. At one point I found myself being screamed at for not getting off an LMTV (a large army truck with a canvas cover) quickly enough and received from the drill sergeant the usual list of reasons I should kill myself. I had started basic with a group suggestion that I be voted for a squad leader position because I was smart. Now, it was automatically assumed I was a slow, weak shitbag. Yet what I didn't know was that my body had contracted pneumonia and I was running a fever of 104 degrees.

It all came to a head one night when I was on my hour of guard and suddenly blacked out, then awoke to find myself surrounded by a group of squad mates, who had pulled me inside the patrol base. After that I was taken to the TOC (Tactical Operations Center) where a group of drill sergeants and medics were set up to support the training. The next morning when I awoke, I was told I was going to be released to my company. I called the number left by Drill Sergeant Fleck and told him the hospital had said it was okay to pick me up. He arrived and we began driving back, but when we arrived and checked at the TOC, I was told I had deserted the hospital and needed to return. The hospital had taken my X-rays and realized the infection had spread through my lungs more than they had believed. I needed antibiotic treatment for a few more days. Meanwhile, my drill sergeant took a vote from the platoon to ask whether I deserved to finish infantry training. Apparently the vote was close, but they voted yes. When I returned, I was not allowed to finish the final march as it had already begun, but I finished training and received my coveted blue cord. I had earned the right to be called Private Russell and begin my career.

I DON'T KNOW BUT I'VE BEEN TOLD,
ESKIMO PUSSY IS MIGHTY COLD
THAW IT OUT AND YOU WILL SEE,
IT'S JUST FINE FOR YOU AND ME!

CHAPTER 4

The Manchu Mile

The next few days were spent preparing for graduation, which is where I began learning that many people did not know how to tie a tie or prepare a dress uniform. I assisted as much as I could, but it was eye-opening and helped me begin to empathize. As orders began to be handed out to individuals, our drill sergeants laughed and told groups of soldiers to get ready to die, because the unit they were being assigned to was deploying to Iraq. Yet I stayed calm, because my contract specified that my first duty station would be in South Korea. I had requested it because my Uncle Paul, son of the famous Pappi Gunn, had fought in the Korean War, and I wanted to see the things he had seen. But mostly I wanted to escape the hellscape of Texas. Drill Sergeant Paul was furious when he found out I was going to Korea. He told me I should be one of the people to die in Iraq, but I was delaying the inevitable. I replied that I had it in my contract. "All right, private, you bought yourself a year," he laughed.

I felt grateful that my parents were going to show up for my graduation. This was the longest I had been on my own, and my parents' support meant a lot to me. Yet when they showed up, they chastised me for begging them to come and said I had drained their savings. In the end they couldn't even take me home. I found out that because of the unit I was assigned to, I had to endure three weeks of Bradley Fighting Vehicle Training before I could leave. My parents looked at me in disgust, as if I was doing this on my own, and told me not to ask for any help getting home. At least I knew I was a shitbag to everyone now.

Drill Sergeant Paul dropped a few privates off at some new units as he drove us to our training facility. I was able to see the introduction of new privates to another unit while waiting in the van. People were hanging off the banisters yelling at the van: "Cherry! Your ass is about to get fucked good!" They were clanking beer bottles over the rails, jumping up and down, whistling, staring at the few privates walking towards the building. It looked like something out of a prison movie or *Mad Max*. We continued on, to be dropped off at the Airborne holdover, and our instructor was meant to meet us there. We were in luck; the course was just a civilian-led informational course. The teacher bought us coffee every day and would drive us to the PX (Post Exchange). The other soldiers in holdover were mostly people who had been injured at Airborne school, or candidates attempting to go through Special Forces Assessment and Selection. This meant we were mostly left alone and unmonitored. I learned that these small respites away from your unit are a godsend. You become self-managed and can act like an adult. You just show up to class, do your work, and leave. What you do in your off time is not monitored, and if you follow class rules and pass, you're left alone. It's a vacation from the bullshit.

Before shipping out to Korea I was given thirty days leave, and the only place I knew to go was back to Houston. I figured I would go there, then spend a few days in Los Angeles before my flight. I spent twenty-five days in Houston and soon realized nothing had changed in my hometown. The few friends I did want to keep up with were acting like deadbeats at their parents' houses, and my best friend James Willkie was beginning college. His parents had given him a car, and he would drive me around to various places, but I didn't feel any different. No one seemed to even care that I had left. Before I knew it, I was on my way to Los Angeles. I stayed in a hotel for a few days and walked on the Hollywood boardwalk and saw the sights and scenes. It was uneventful, and soon I was on a flight to Korea.

I began the flight on my own, disappointed that I hadn't experienced any adventure in Hollywood. I hadn't seen any stars, I couldn't drink, and I certainly didn't hook up. I was almost too ready to leave when I did. As I boarded the flight, I made sure I was not in uniform and tried to maintain a low profile. Some asshats were wearing their uniforms so they could board the plane first and have every fuckface in line thank them for their service. Many POGs revel in this behavior and can't wait for someone to fill the void in their soul by getting validation and praise from random strangers. Infantrymen don't do this. You're there for one reason, and that's the pink mist. I sat in the back trying to stay out of sight, but I did get to trade for the window seat. Three guys from my basic training boarded as we connected in Washington State. I tried to lay low, but eventually they noticed me and traded some seats around so we could sit together. I was trying to hang onto the last vestige of individuality I could feel. Every time you get around the brothers, your privacy ceases, and your sense of self fades away.

By the time we arrived in Korea, everyone had swapped stories of leave and anticipated the new country we were about to arrive in. The plane touched down, and we grabbed our bags. As I passed through the door, I noticed there wasn't a connecting ramp from the tarmac to the terminal. I walked outside and froze, literally. I had left Texas and California at nearly 85 degrees in November. In Korea it was just above zero and snowing. I remembered stories from my uncle about the bitter cold in Korea, how it froze your body to the bone, and at this point in my life I had seen snow twice. I was pushed forward by the person behind me. The faster I tried to move, the stiffer my body felt and the harder it was to breathe.

We sat through some mandatory briefings, got a schedule, and went from there. More inoculations, paperwork drama, and they issued your gear. This all happened over a week, and I felt like I was back in 30th AG again. I was starting to see a pattern here. (The military becomes one familiar cycle that just repeats itself every time you change duty stations.) Yet while you wait you meet more people you knew from various parts of the Army. A few more people from basic arrived, and we all stayed together. Our group started out at least twenty-five strong, but by the end of the week our number dwindled to fourteen. Every day after you finish your in-processing requirements you show up to a lobby, almost like an aircraft terminal with wooden benches, and you wait. By now I was beginning to get a sense of my environment and taking note of things around me. The only place we could go was across the street for our CIF (Central Issue Facility) issue of our gear or to eat at the only restaurant on base. They served three different noodle bowls of *bulgogi* and a type of chicken. We were told tales of how harsh the country was, and the rules and regulations we were required to live under. It seemed like an open-air prison. You couldn't stay out after 2200, leave town without a pass, go anywhere alone. One sergeant joked while briefing us that the fastest way to leave Korea as an E5 (SGT) was to arrive as an E6 (SSG). We laughed, but little did I know how true it was. The drinking age was also restricted to U.S. law, which meant twenty-one, and military police would walk in public and check your ID at any time. The reason the rules were so strict was that the United States was still at war with North Korea. You might hear that the longest war we were involved in was Vietnam or Afghanistan, but most people don't know that even though conflict ended in Korea in 1953, we never signed an armistice, only a ceasefire. Technically, my first duty station was a war zone, it was just inactive. This place was a limbo like no other, but a trap for every twenty-year-old male seeking adventure. It was Shangri-La and Tartarus at the same time.

One morning we arrived at the terminal assuming we were going to sit and wait again, but the E7 (SFC) had become tired of our antics and told us we were going to be assigned to a POG unit because all the Infantry units were full up. I was relieved at the thought of not having to join an Infantry unit immediately and was plotting ways to explore the countryside on my own. Yet for now it seemed we were stuck. We were told that the first thirty days in country we were required to be on lockdown, which meant we couldn't leave our unit or base. We had to be monitored and check in regularly. Yet at a stroke of luck, we found out a small Airborne Infantry unit was available to take some new privates. It was in Uijongbu, and the SFC told us some of the rules were a little laxer there. I began to feel more optimistic.

This was something I would learn to regret, then after a while give up altogether. In the Army you must keep your expectations low. If you can imagine every scenario as ending in the worst possible outcome, you will start to be pleasantly surprised. I found out that the unit was an Airborne unit, so I bought some jump boots, a 2ID unit patch, and a maroon beret. Although you want to show up to a brand-new unit as squared away as possible, you don't want to wear or display things that you haven't earned. I wasn't Airborne qualified and didn't deserve to wear jump boots or a maroon beret. I think back to a few years ago when there were swaths of parents calling unit commanders demanding their pussy of a son or daughter be allowed to wear a combat patch they hadn't earned the right to wear. I didn't have that same intent; I just had been told that Airborne units wear these things and thought I should be squared away. The sentiment didn't go as planned, and in retrospect I agree with the anger and retribution that came my way. We were picked up by a corporal (E4), and if you know anything about the rank structure in the Army, these people are notorious for being the biggest dicks in the military. On paper a corporal has the authority of a sergeant, but really only as much as the other noncommissioned officers allow. They have more responsibility but receive no extra pay or benefits. I have also witnessed many corporals stay in this rank much longer than other E4s that are SPC. This particular corporal fit the stereotype. He sternly told us to get our shit and meet him outside.

One of the privates in the group started to ask a question, and the corporal cut him off: "Did I fucking stutter?" Then, as we began to walk away, he calmly asked, "Do all of you fucking move like old people fuck?"

We ran back to the temporary barracks, then threw our shit out into the truck. Noticing that I was wearing a maroon beret that wasn't even shaven or trimmed, and jump boots that weren't shined correctly, the corporal asked me what company I had been in during jump school. I told him I hadn't attended Airborne. His eyes lit up with such joy and anger at the same time. He told me I was going to get skull fucked into the dirt, and to get into the truck. Once inside the back of the LMTV (a large army truck with a canvas cover), he turned back to us and informed us that the unit hadn't had privates arrive in over a year, and they were waiting to pop cherries. The drive to Camp Essayons was the longest ride of my life. I didn't know what I had done wrong, but it was apparent that wearing the beret and boots was some sort of disrespect that wasn't tolerated. I changed out of them in the back of the truck.

We went through the gate, and I noticed it was covered in a green tarp that read KNOWLEDGE FOR BATTLE. There was a sour, stale smell in the air, and the grayness of the ambiance had almost intensified. Soon everything would begin moving faster. The corporal asked for our ID cards and began screaming almost the second after we failed to immediately give them to him. We heard a commotion from the gate guards, but I couldn't make out what they were saying. I did notice the corporal laughing, and my anxiety increased. I could feel myself shaking, my heart was pounding, and then the truck stopped. We were told to get out. I was in the back. Some of the guys started grabbing their bags, until the corporal yelled that he had not specifically told them to do that, but to just get the fuck out of the truck. He snarled that we were all retarded but were about to get ours. He then told us to form up, but no matter how we formed up we were wrong. We made one line, and were yelled at to create two, then we created two ranks (horizontal lines) and were yelled at that we needed to form files (vertical lines). Once we formed two files, they said that was stupid and to form ranks again.

Eventually the corporal came over to me and noticed I had changed. He looked back at the other soldiers watching from afar and screamed: "Who was the fucking dumb dick who thought he should wear a beret and jump boots even though he's not Airborne?"

I said it was me. He reminded me that I was supposed to add "corporal" after his name. He then forced me to go back to the truck and put my boots and beret back on.

"I want you to look really good when your new squad leader fucks the shit out of you, private!"

I knew I had fucked up. I ran to the truck and began digging through my bags. I flung the boots on and barely tied them together. I put my black beret in my pocket and ran out the back of the truck just in time to hear the corporal yelling that it was time to go. We were taken inside to a conference room with a round table and told to enter. One private began to sit down and was immediately snapped at to stand until told to sit. We waited, our anxiety flooding the room and bouncing off one another. We were like fish who can see the shark circling our school but don't yet know which one will be singled out. The CO walked in, and we were called to attention. He told us "At ease" and encouraged us to sit, then went over the base and explained that we were in an MI (Military Intelligence) battalion and our job was to gather INTEL (intelligence) in the field for the battalion to take back to division so that they could prepare missions and training. This was a premier unit with Green Berets and Rangers, and Airborne qualified except obviously for us. We were given information on the battalion commander, a female lieutenant colonel.

After a long, drawn-out introduction, we were released to the NCOs (non-commissioned officers) who were supposed to divvy us up like playing cards. It was an eerie feeling knowing that if you were picked last, you appeared weak, like getting picked last for kickball. If you were picked first, you were spotlighted and had some level of expectations. As the unusually large men towered over us, they began to pick. Lucky for me I was picked third or fourth out of fourteen. I was hoping this would allow me to lay low and not get fucked with. The CO then informed us the brief was over.

"I'm sure you've all been through what's about to happen next," he added.

He left the room and we were called to attention, then told at ease. Then the screaming began. The shouting to run outside and wait was repeated until every man was outside. Once again, the rule was: Don't be last. The group of NCOs began with a light round of pushups, flutter kicks, and other exercises to start the hazing with. Then they brought out a dog that looked like a giant chow, and it began snarling and barking at us. One of the NCOs remarked that the dog hated privates. When the dog went to like one of us, we were all furiously forced to complete more labors while the other men screamed that we weren't worthy of touching the dog. Her name was Momma, and she outranked every person in the company. We had to stand at attention and ask her if she wanted us to pet her. This show of force and aggression only riled the dog up, and she began to growl and bark louder. We were once again commanded to form up, and even though we started to do so like before, we were yelled at that we were in new squads and needed to divide up just like that. Then it changed again, as the different squad leaders continued trading us like cards. They noticed who had been weaker in some areas, while others had better heart than others. We were once again divided into groups of three. Now I noticed two other guys next to me. Woopie and Tad were both from my basic training platoon. Our new squad leader came over and began to mean mug us with a smirk. He was slightly portly, but still muscular, and his hair was blond almost to the point of being white.

His name was SSG Schaefer, and even though he would be my squad leader for less than a month, he made a significant impact on my arrival to Korea. He began immediately looking at my maroon beret and glossy jump boots. He inquired about my company at jump school, and once he found out I had not been, it was over. I was stealing valor and taking credit for achievements that were not mine. We were forced to run all over base while stopping to get smoked at various buildings of repute, such as the shoppette, PX, library, and chow hall. Each time we stopped we were reminded that we were scum and did not deserve to arrive at this base. It was already sickening enough that a liar such as myself should be parading around with jump boots and a maroon beret, but I would pay with my sweat for the mistake, and so would the two new privates in my squad. This also meant the "Don't be last" rule didn't apply anymore. I was beginning to buy the indoctrination that I was a liar who was parading around with jump boots and a maroon beret. Yet what bothered me more was that my two teammates were paying for my mistake.

This didn't stop me from trying my best. I tried to show no pain, weakness, or fatigue. I always responded with "YES, Sergeant!" and would never admit I was tired. By the time the tour was over our new squad leader told me to take off my fucking beret and give it to him. He threw it in the garbage and told me to get a black beret and forget about ever being fucking Airborne.

Honestly, I didn't care. I just wanted this to end. We were forced to low crawl across a volleyball field and sprint to our barracks. We were then told some ground rules about not being able to leave the base for the weekend, and under no circumstance could we even look at alcohol. We were then taken to our barracks rooms, which were two long Quonset huts connected by a public shower/bathroom/laundry room. It looked like the building was built during the Korean War, and I remember our CO telling us that the temporary living quarters was, for lack of a better word, condemned. It certainly lived up to expectations: holes in the walls, mold in the corners, and it smelled like old piss and socks caked in cum. We were given our room key and left alone, but soon enough two other very large men walked in. Specialists Matt and Dinoyo were our new team leaders who were going to show us the ropes. Matt looked like an ugly Channing Tatum with pockmarks and scars, and Dinoyo was a large Samoan with hair out of regulation and full sleeve tattoos. They told us the brutality had ended for the weekend, then spent a few hours talking to us and helping us set up our gear. They showed us various knots to use to tie down our rucksacks and LBV (load bearing vest to hold ammunition). They told us we were in a special unit in the Army called LRS-D, or Long-Range Surveillance Detachment, and our mission in a combat scenario would be to gather intelligence on enemy units, then possibly provide direct action before a main body or larger unit would arrive. We were lucky because many people in the Army wanted to be in these types of units, and we had accidentally been assigned. SPC Dinoyo quipped that our entire time in the Army would be ruined because we had started in such a unit, and I didn't realize what he meant by that, but later I would.

We talked some more about where we came from, and joked a bit, but they finally laid out the expectation. The next week, after we finished in-processing the unit, we would have to go through the LRS assessment. We would be tested on our knowledge, skills, and physical fitness. We would have to conduct an official PT test, a land navigation course, one night in the field where we conducted nighttime land navigation and showed we knew how to set up a radio, then a twelve-mile ruck march. The last part would be a swim test where we would be pushed in the water with a rucksack, our weapon, and full gear and boots, and must tread water for fifteen minutes. It was all very daunting, and I wondered if I could do it. Our longest march at basic was a twelve-mile at the end, and it nearly killed me. I no longer trusted myself after my fuckup with the boots and beret. In the meantime they told us to rest, and they would grab us in the morning. They planned to take us to breakfast across the street from the base. I reminded them that the squad leaders had told us we would be punished if we left the base, and they told us it was okay if we were with them. We just couldn't go out alone without someone who had been in country less than six months. We were never allowed to go anywhere alone. I set up my bed, and Tad was my roommate. The other guy, Woopie, had been paired in another room across the common area. We began to rack out and ready ourselves for the day to come, when Tad suddenly spoke up. He said he heard whispering, and I told him it was the wind. My eyes were heavy. I fell asleep.

Dawn peeked in through the sad single-pane window above my bed. I tried to roll over, but Tad finally got up and began rummaging through his duffel bag. He asked if I was awake, and I sadly responded yes. He told me he hadn't slept well and had been hearing whispering almost all night long. I listened and told him I didn't hear anything, and he told me it had stopped a few hours earlier. We began looking around and checking the two windows to make sure they weren't open. After that I asked if he knew where it came from, and he told me he had heard it behind his dresser. We pulled the dresser back, and there was a goddamn face in the fucking wall. As soon as the dresser pulled back, the face opened its eyes and screamed at us. We fell backwards and could hear laughter on the other side of the wall. The man in the wall walked around and knocked on our door. I don't remember his name, but he was from the MI (military intelligence) side of the BN (battalion) and told us he had been waiting all night for us to find him to scare us. He had drunk three bottles of bourbon and pissed himself, but it was worth it. We didn't know what to make of it, but he marched off to the bathroom, throwing off his clothes in the hallway as he walked. What the fuck had I been thinking when I signed a legal fucking contract to owe my life to the government for eight years of my life?

The next week was a blur consisting of the normal indoctrination infantry privates receive on arrival at a base. Usually after your first hazing you aren't smoked until your first PT test. We waited in line for paperwork and had small packets of information containing basic Army and Infantry knowledge we were meant to study constantly. We were verbally corrected for uniform infractions such as boots not being shined or BDUs not pressed well enough, but everything was calm and moving towards the assessment. We were taught how to pack our gear and clothes into our rucks, and how to use the wet weather bag to seal the inside of the ruck to make a flotation device. We were given basic map-reading classes and other informative classes on LRS-D operations. Yet the day arrived when we would be taking the LRS-D Assessment and tested against all odds.

Most of the fourteen of us were pretty sure we could handle the assessment, except me and Bruce. We decided to stick together. I told him my idea of not being first, but not being last either. He said he would literally come in last for me if he had to, because he didn't really care either way. I couldn't believe someone would do that, but there it was. The various tests weren't as difficult as I had thought they would be, until the nighttime land navigation course. We would be performing it with no Night Vision or NODs and were supposed to rely on a compass and red light. I looked at Bruce and asked how his Land Nav skills were, and he said he had done well in basic. We teamed up and got the same scorecard and barely made it through on time. One team didn't make it. They came back arguing and were told to cut it out.

Now it was time to sleep in the woods and make the twelve-mile trek home. I didn't know how they would route it since we were only five minutes from base, but the route would be an urban one through Uijongbu. We got the experience of ruck marching while seeing the sights and local people. I began waving at some and was immediately shut down. Apparently looking friendly in uniform was a social crime in this unit. I just kept going, and when I thought my feet couldn't take it anymore, we were back at base. What I didn't realize was that I hadn't remembered seeing any water we could float in on base. We marched back to base to find a pool on post and stood at the edge while we were given ten minutes to drink water. We couldn't drop our gear or rucks, and I drank a very small amount of water from my canteen, as I knew they could also help with flotation. I had adjusted the canteens to be square on my hip bones, below my ALICE (all-purpose lightweight individual carrying equipment) pack, which would help lift the metal frame from my hips a bit. SPCs Matt and Dinovo had taught our team how to pack the ruck with the wet weather bag to also have it work as a flotation device.

When it was time, we jumped in the pool and were told to hold our rifles above our heads and not let them get wet. Many began furiously kicking, and Bruce looked panicked. I told him to look at how I was kicking my legs and match my rhythm. I grew up on the water and swam quite a bit. I knew that you can't fight the water, you must move with it. We had packed our rucks just right so that we could lean back on them occasionally, and just as they felt like they were sinking we began kicking again. The sergeants all began yelling how much time had passed and slurring to intimidate us or make us want to quit. I began kicking my legs out to the side and resting. The water cooled my body, and I felt very relaxed. One man, however, hadn't packed his ruck correctly, and it broke and filled with water. We had five minutes left. They told him if he couldn't make the time, they would pick his dead ass out of the water, resuscitate him, throw his ass back in, and restart his time. He sank and lost the fight. They pulled his ass out and surely enough threw him back in. We all gained a little more motivation. After the second time he sank they pulled him out and told him to wait outside. He had failed this portion of the assessment. Bruce and I continued swimming and made it to the end.

By the time we got out of the water, I wondered what I had been so afraid of. Yet as we were being congratulated on our success, the bidding war for privates began again. Some squad leaders had buyer's remorse. One said he had lost a private completely; the guy who failed the swim test also failed the land navigation and some other test. They laughed at us and said we were the unlucky thirteen. We got the rest of the day off and were told we would be split up once more in the morning.

OLD KING COLE WAS A MERRY OLD SOUL
AND A MERRY OLD SOUL WAS HE.
HE CALLED FOR HIS PIPE
AND CALLED FOR HIS BOWEL
AND HE CALLED FOR HIS PRIVATES THREE.

BEER BEER BEER SAID THE PRIVATES.
MERRY MEN ARE WE!

BUT NONE SO FAIR THAT WE CAN COMPARE
TO THE AIRBORNE INFANTRY!

CHAPTER 5

LRS-D

The next morning the entire company was present for formation. The one man who hadn't made it through the assessment was not. He would get moved to the post mailroom and technically be on the books as part of the company, but he was not LRS. Apparently, it wasn't just us privates who were going to get shuffled around; the entire company was getting a makeover. A large enough number of soldiers were PCS/ETSing (leaving to another duty station/leaving the army), so it made sense to restructure the company. My squad leader was one who was leaving so Tate, Bruce, and I stayed in the same squad but received a new squad leader. SGT Farmer and SPC Matt were the team leaders.

SSG Dilbeck was the scariest human being on the planet. He stood at least 6'3" and was a solid 250 pounds. Everyone feared SSG Dilbeck – our platoon sergeant, first sergeant, everyone. His face was constantly angry, and he could chug an entire bottle of Grey Goose in one chug. The man had his demons and drank with them regularly. My first day in his squad, he came to my room after work, kicked in my door, and demanded I drink with him. It was a Sunday, and we had to be up early as usual for Monday morning PT. He asked if I had ever shown up to formation drunk, and before I could verbalize that I would never do that, he threw down a case of Bud Light. He showed me the straw trick, and after six I was wasted. I threw up that night, and he nearly fell off my balcony. He would gather the squad each night before we got recalled for the active alerts in Korea. They would test our response time to see how quickly we could react if North Korea were to attack. He would make sure we were ready, then we would spend the night drinking. The next morning, after showing up ready to go and drawing our weapons from the arms room, he would usher us back to his room and have the medic give us IVs to cure the alcohol and hangover. He once walked into the Stalker club on base and was hit in the head with a dart because the board was on the wall next to the door. The whole club stared in silence as he easily pulled it out, then announced it was time to start drinking. He would mess with us constantly, but one day he demanded I get him a box of grid squares. I knew they didn't exist, but I wanted to beat him at his own game. I went to the Company building, grabbed a map of our base, cut out the grid squares, stuffed them in a shoebox, and waited all day in my room. When I didn't show up to final formation, he came to my room extremely pissed, asking where I had been. I informed him that I had found the box of grid squares, and he slowly realized where I had gotten them from. He went into a complete rage, screaming and cursing. He smoked me all the way up to the Company and the 1SG came out to see what was happening. Seeing me in a full sweat, he asked SSG Dilbeck what I had done.

"This fucking private cut up the company map!!!"

"So, you told your private to get you a box of grid squares, and he did it, so now you're smoking him?"

SSG Dilbeck's face lost color as he realized what was about to happen. 1SG then told me to leave, and I had the rest of the day off. He then told my squad leader to stay and tape the map back together, grid square by grid square. The next day I stayed in my room, terrified of what SSG Dilbeck was going to do to me. However, when I showed up to PT the following day, he told me: "Touché, Russell, but you'll never get another one over me."

"Roger sergeant," and I smiled.

But not all was brutal hazing and what we liked to call fuck-fuck games.

One night, Papa Roach came to play at Camp Essayons. After the concert he went to the only bar on base, the Stalker Club. He sat down with my squad and was asking us about the show. I told him I didn't really like his music, and he told me it didn't matter, he was a millionaire. He then proceeded to buy drinks for the entire bar. There were many other nights of drinking, partying, and revelry. We would get into taxis and play a game called Odeshi racing. Each person would offer the driver five extra dollars if he beat the other cab. They would jump curbs, blow through lights, and drive recklessly. I still gave the cab driver the money, even if he didn't win. We would go out as a full squad to Seoul and drink until we couldn't see straight. We ruck marched there a few times with liquor in our rucks, then stopped at a hotel before heading out and drinking the night away. I remember going to a bar called the Honkey Tonk that played nothing but gangsta rap and had a free beer tap on the wall. We went to Korean karaoke bars, signified by a large, wooden door sporting demon faces. We drank soju bowls and snuck our way into clubs that required you to be twenty-one.

One time we went as a company to a local bar and began drinking after a field problem where we successfully played wargames against our MI sister company. We went to this bar because in Korea, if you put a $20 bill on the bar and it had your chain of command's names on it, it was "your" bar, meaning you got free or reduced drinks. However, two guys from a neighboring base came by and thought they would place their bill over ours. What they didn't know was that our entire company was in the bar. They must have had a few friends, because when an argument erupted a few others stood up to calm the situation. One of the guys began calling out a guy in our company named Wallace, a small, redheaded bastard with yellow eyes. Wallace's temper was extreme, and there were stories that he had deployed with 2nd Ranger Battalion in the initial invasion and had gunned down all sorts of women and children in Fallujah. I never knew if it was hype to scare us, or real, but his temper was something I wouldn't test. As the man began to sling his shit at him, Wallace ran to the DJ table and flew across the turntables and knocked out the DJ. The bar exploded into a full brawl. We were throwing bottles and chairs and hitting anyone that came near. A large Ranger named Ford was literally picking up people with one hand, hitting them, then tossing them like a toy. Once we had efficiently damaged the establishment, we ran back to base, and our company commander, who had been there, told us no man was to speak of this ever again.

We had to work just as hard as we played. About our critics our female battalion Commander often said, "Boys will be boys." Things like blowing up a propane tank on the hill we called the Hindsight were met with physical punishments rather than any legal or non-judicial actions. The only respite I would get from this insanity was meeting my cousin, who had brought his wife to Korea and lived off post. He let me stay a few weekends at his house at Camp Stanley. We would go out and drink, and I noticed the non-combat MOSes had some things in common that we, the grunts, did not. This was not a livelihood for them, this was a job. They didn't care about the Army culture or value the lifestyle that supported them, and most of all they did not respect authority. I was regaled with tales of how to fuck over your senior enlisted NCOs and how to get what you wanted out of the Army. I listened to them, time and time again, say they wouldn't put up with the shit I put up with. Yet, they didn't know SGT Dillbeck. They also didn't know the reward of belonging to the brotherhood. Reality would hit when I went back to work. They were POGs, and they didn't live in my world.

Soon our company was informed that we would be taking a tour of the JSA (joint security area), the joint base on the DMZ (demilitarized zone). It was like a tourist event; they showed us the three blue houses where meetings would occur between the two Koreas. The two houses on the side were called the monkey houses because during the meetings the North Korean soldiers would fill them up and begin shouting, pulling out knives, imitating slitting their throats, and jumping around. The South Korean soldiers would have to stand in the modified Tai Kwon Do stance and only silhouette half of their body for their twelve-hour guard shifts. They looked ridiculous with their green outfits, black UN helmets, and aviator glasses. You were not allowed to point across the DMZ, or the North Koreans would shoot at you. We were shown the lone road crossing the DMZ and told of the two people who had made it across. One was a Russian spy who ran across the road while being shot at. The other was an old man who would pretend to paint but was mapping out where animals would not be harmed by land mines. He crossed under the cover of night and was given amnesty. There were communities of North Koreans who were being paid to stay and farm on the DMZ on the southern side. They had been caught on that side after the ceasefire and were not allowed to leave.

The weirdest thing was the city on the other side of the DMZ. It was completely white and had a giant flag on top of the tallest building. It was all a show. The city was made of plywood and constantly blasted propaganda about how great North Korea was, how if you went to their side they would bless you, and so on. We were told there was a million-dollar bounty for one square foot of the flag. During the 1980s a group of soldiers had been hacked to death with axes by the North Koreans for trying to chop down a cherry tree. Technically there was only a ceasefire agreement, and the Korean conflict was the longest war the U.S. has ever been engaged in. Standing in the blue houses, we could go to the other side of the room and were told we were technically in North Korea.

At the end of the tour we were told that we would soon begin our tour at the JSA to pull security. We did a month tour along the DMZ, and each day was as ridiculous as the last. I had to walk a portion of the DMZ as a North Korean walked the opposite side. We were told not to look at or talk to the North Korean soldiers at all. Of course I didn't listen and tried to make idle conversation. The soldiers immediately began jumping up and down and making threats with their knives. I was severely reprimanded for that.

After our stint at the DMZ, we were allowed to take leave. Everyone else wanted to go home to the U.S. I wanted time alone in Korea. The one thing I wanted was the ability to walk around alone. So, for a week I backpacked around from Seoul to Pusan. I visited Buddhist sites and war memorials but kept my leave form close, in case I was stopped by MPs. One day while walking through a small town, I decided to eat dinner at a local restaurant. I sat down and watched a woman come out to serve me, and she began putting out the usual condiments and prepared some tea. Normally in Korea a long table will be full of lettuce leaves, bean sprouts, different types of kimchi, and assorted vegetables. This time, something unusual was placed in front of me. It was a small bowl, and it spilled some water as the momma put it on the table. I thought it was an accident, so I looked up to see what had been spilled. A slim black creature lay inside what looked like a white ceramic ashtray of water. It was squirming and moving. I looked up in horror, only to realize that a woman sitting across from me was eating these creatures with chopsticks like candy. Just as she slurped one up, I could tell what I was looking at: a baby octopus. The woman looked like a teenager, and she was just fisting these gross creatures with her chopsticks. I couldn't be shown up. I had to eat this thing. I then noticed the entire restaurant was staring at me. Was the white boy going to eat the live octopus? You're damn right I was.

I started fighting with the thing using my chopsticks, and it kept squirming away and inking in the bowl. I couldn't get a grip. Just before I was about to say screw it and use my hands, I finally pinned it and picked it up. I didn't want to drop it, so I threw it quickly into my mouth. As it began squirming all over my tongue, I realized something. It had suckers on its tentacles, and what if it decided to stick to my throat? It was too late now. The taste of snot and salt was filling my mouth. I'm pretty sure there was a metallic taste to the ink as well. I swallowed the octopus whole, and the crowd went wild! They all cheered, and the momma gave me some soju. She asked, "Baby, you want another?" I declined.

LRS-D would eventually be disbanded. This meant that everyone in the company would be sent to different bases and units throughout the country. Our last mission was to conduct FORSCOM security for major bases in Korea. We were selected to guard Camp Red Cloud, the divisional headquarters. South Korea requires its citizens to register in order to protest. They must tell the government when and where they plan to do their actions. Most of the time, they would protest at the Uijonbu train station. We just avoided the station, but before I had arrived in the country two girls had been run over by Bradley fighting vehicles when they tried to cross the street. It seemed to be a complete accident and misunderstanding of cultural norms. Koreans often just raise their hand in the air and walk across streets and expect traffic to move around them, even on busy streets. A hundred-ton tracked vehicle cannot stop on a dime and cannot maneuver like a car. The girls apparently walked out of a low-lying ditch and were run over. This led to a giant protest scheduled outside Camp Red Cloud, to demand action from the U.S. military for those soldiers involved. No one knew it would get out of hand, and we had taken some riot classes from the MPs and knew how to put on the gear and use the mace we had been issued. They had given us leather police tools called blackjacks, much like batons, but took them away once we began to use them on each other. The day of the riot we were already in gear, ready to react, and then we were called down to the gate. There was a lot of commotion, and the Korean police were already yelling over bullhorns and firing teargas into the crowd. Two men came forward with a propane tank on a pole, lit it on fire, and moved forward in the crowd, carrying the spinning fire wheel. It didn't take long for the people to overpower the police and move past the blockade of police and buses. They began cutting the chain link fence and moving into the base. The first line soldiers clashed with the people, and they were bashing against the shields. I saw another guy from a different PLT get hit in the face, and his nose was bleeding. About that time my squad leader noticed people jumping the fence from trees and took our squad over to detain them. I was happy I didn't have to go to the front line that day. The civilians left once they realized they couldn't push us back any further. It finished almost as soon as it began, and we were left wondering what to do with all the people we had detained. Now I had a different idea of what my service entailed. During the riot I had been terrified, but I hadn't run. Would I be able to endure battle or a firefight? We went back to our base, and our disbandment was made official.

The last day of our unit we got together and had a party for the ages. People were getting drunk all over the place, in barracks, in the motor pool, even in the company building. We had worked all week to turn our company property over to the supply sergeants who would stay behind and physically transfer it. When I woke up in the morning, the sun was shining and wind was blowing against my face. I had passed out in the front yard of the barracks. I noticed the battalion commander stepping over me. She looked down, shushed me with her finger, and told me to go back to sleep. I passed out again but was soon woken up for the punishment that would ensue. Apparently two lieutenants in their drunken rage had thrown beer bottles at Humvees until the windows cracked. All the officers were relieved of their duties, and we were required to stay in our day room for 24 hours while being monitored. We all felt like shit, but the last huzzah was worth it.

WAY DOWN IN THE VALLEY
I HEARD A MIGHTY ROAR.
IT WAS THE SOUND OF EVIL ECHO
USING CHARLIE AS A TOY
PUT YOUR FOOT ON THE PEDAL!
PRESS HARD ON THE GAS!
MOVE OVER CHICKEN CHARLIE,
AND LET EVIL ECHO PASS!

CHAPTER 6

Dragon Valley

Within the week we would get transferred to Camp Casey, to a place called Dragon Valley at the very last stop on the base. I had heard rumors about the place; supposedly this is where the "Big Army" and tightwads lived. We were warned that this was where everyone got issued their Article 15s en masse.

Camp Casey sucked. People were living on top of each other in a community that was always on edge. Dragon Hill was the place the busses stopped servicing first and started last, so when you got back from the front gate it was rare that a bus would be running to take you up there. Most of the time we would end up walking the long road of shame for two miles. This would also be my first interaction with PSYOPS, or psychological operations, the Army's version of the CIA, professional liars. My friends and I were drinking in a bar, and a girl was trying to act like a drinky girl to help CID (Criminal Investigation Department) arrest male soldiers trying to pay for sex. Drinky girls would have you pay for drinks and pretend to be your girlfriend, and many soldiers spent their entire paychecks on these women. What made the practice a target was that most drinky bars would go further by having the soldier buy the girl's night off and take her to a hotel for sex. Yet the blonde white girl was immediately found out because a couple of guys from my platoon got her drunk and she blew her cover. Yet I would find my own trouble by trying to avoid curfew one night. I figured I had done it enough without getting caught that it wouldn't be a big deal. I showed back up Sunday and saw my entire PLT in formation and in uniform. I tried to slip through the side doors, but they were guarded. So I walked up front to see my chain of command from my team leader up in the front area with CQ (Charge of Quarters). Apparently another soldier had gone AWOL, forged the CO's signature, and left the country. We had a full recall and, since I was getting blitzed in Seoul, I hadn't known about it. They asked where I had been and, when I told them, they flipped their shit. You can't leave the city without a pass. I got completely hemmed up and ended up with a field grade Article 15: loss of rank and pay and thirty days extra duty. I spent my first eighteen months in the Army busting for early promotion, only to get knocked down to E1 on month 17.

I couldn't believe the audacity of the people on extra duty. Most were drug offenders, some had assault charges or worse. These people would complain about getting extra duty for hitting an officer or NCO or getting caught with heroin. It amazed me that they took no responsibility for what they had done. In the real world, they would have been serving hard time in prison. Here the worst consequence would be getting kicked out of the Army, and even then, in most cases they could keep their benefits. And even still I was a knucklehead. I fell into their line of thinking and cursed the NCOs above me. I shook my fist at authority and spat at those in leadership positions. There was no reason to do this. I had made my bed. I should have kept my head down and complied with my punishment. Instead, I smuggled beer onto my extra duty shifts and drank it during work. I worked as half assed as possible and did the bare minimum to stay under the radar. Back at the barracks I drank with a few private buddies and woke up to do it again. I hadn't learned my lesson.

As the time grew near for me to leave Korea, my friends began leaving. They were coming down with orders and finding their new assignments. Every one of them was off to Iraq. They would head out of Korea as SPC/E4s. I was the only one in that initial briefing room in Osan that was walking out an E1. I was a fucking moron, yet I thought the world of myself, as morons do. I became a recluse and would order in food. I played video games and chatted online. I struck up a conversation with a girl from high school. Soon the conversation was turning to marriage. I began to imagine what would have happened if I hadn't let go of my high school crush, Jennifer Lemus. I began hating myself and regretting everything. Yet I had one more embarrassment to endure before this fucking hellhole would let me go.

I was mopping the CQ one day, waiting for my new orders, and out of the blue I heard my name. "Fucking Russell…" There it was, that voice, the tenor, the gravitas, the gangster. I dropped the mop. I had just heard Drill Sergeant Paul, but there was no way that when I turned around it would be the same man. I would wake up at any second from a drug- or booze-induced state and find out I had missed formation, gone AWOL. Hell, I'd wake up and find out I was being accused of murder. Any of these outcomes was better than turning around and seeing my drill sergeant. I turned, and there he was. Everything except the brown round. I shakily said, "Drill Sergeant Paul…"

"Fuck you Pri, I ain't no drill sergeant anymore. You use my fucking rank, shitbag!"

"Yes, Sergeant…. First…Class…"

"Just say fucking Sergeant, RETARD!"

At this point people were snickering.

"Sergeant…"

"Shut the fuck up, Pri! I told you before you left basic, and I fucking called it. You weren't ever going to be shit. Here you are, E1, fucking shitbag!"

The guys in the room were openly laughing. Some looked like they were about to join in.

"Shut the fuck up, the rest of ya! The rest of yous look like fucking weak sauce shitbags too. Fucking shit sandwiches."

Hell opened and the void took me under. I was on my last field exercise (FTX) and had been placed on ammo detail. I wasn't doing much of anything important for the exercise itself because I was leaving. The sergeant in charge of me was about fed up with my bullshit, and even though he was a very chill person, he decided I needed to be punished for running my mouth. He told me to start pushing, and I refused. I held my ground because for some unknown reason I felt it was important. Drill Sergeant Paul noted this going on and stepped in. He said he was the only person that could inspire fear in me and would make me comply. Now, I had to stand my ground. I had to show I wasn't afraid. I had bluffed, and I had to come through. He started screaming, yet I refused to back down. He finally threatened to give me another Article 15 and keep me in country longer. I complied with his demand and began pushing. This was stupid, but at the time my self-pitying anger was more about showing I wasn't afraid or weak than about defiance. I just wanted to prove I wasn't afraid anymore. At twenty, who really makes the right decisions on any of this? Certainly not me.

Afterwards I was let go and left the country. Unlike other people, I didn't have a group of friends to help me to the bus in the morning. I didn't have a sergeant willing to help with the clearing process. I was on my own, and this is what I wanted. In almost two years in the Army, I hadn't had any real autonomy or privacy. The constant strain of dealing with people was gone. I went to a Drowning Pool concert the night before I left and drank myself blackout drunk. I woke up barely in time to crawl down the hill to the bus. I busted my suitcase, lost a wheel, and dragged the thing until it ripped open at three a.m. At the airport I pulled out my 100MPH tape and did my best to patch the hole, but it didn't hold together. By the time I arrived in the U.S. the whole case was nearly open. My dad met me at the airport and brought me back to the house, and all was well for a few minutes. I tried hooking up with the girl I had been speaking with online, and it fell apart. Most of my friends couldn't see me for this reason or that. I would find that to be a common occurrence after Korea. On top of this my mother didn't like that I was in the house. She was insistent that I should leave and go on a vacation, and any complaint about the Army was met with: "Well, you should have joined the Air Force or gone to college."

I was shunned by the brotherhood in the Army and shunned at home. This time in between my orders to Fort Hood, Texas allowed me to reflect on my attitude, and I was embarrassed. I just had to change who I was. The last two years I had been fighting, and for what? Was I supposed to hold onto a piece of myself that wasn't working? I couldn't move forward, and feel accepted, if I didn't change. I decided it was time to head to Fort Hood with a new attitude, and I left leave early, catching a Greyhound bus to Killeen.

PART TWO

SITTING ON A MOUNTAINTOP BEATING MY DRUM
I BEAT SO LOUD THAT THE MPS COME.
I SAID WIMPY WIMPY DON'T ARREST ME!
ARREST THAT POG BEHIND THAT TREE
HE STOLE THE WHISKEY, I STOLE THE WINE!
NOW ALL I DO IS DOUBLE TIME

CHAPTER 7

Fort Hood

At Fort Hood I hit the ground running, determined to change my outcomes in life. I was ready to embrace the life I had chosen. I laughed at POGs at Fort Hood and began yelling boisterously. I was a class clown after all, and it was easy to be an ass, yet there was no more timidness or insecurity left. I was here to drink, fuck, and fight, and nothing would get in my way. It also didn't hurt that no one from Korea followed me to my new unit. I needed a fresh start, and here it was. I wasn't about to waste it.

I eventually was picked up by the man who would be my PSG, SSG Fili. He was a large Samoan who had more of a model look than an Army look. He didn't have the usual high and tight haircut; he obviously took care to groom himself beyond standard. Also, I'm pretty sure he was wearing cologne. He walked with a swagger and wore his BDUs loose. You could see the sag in his pants, and his trousers came much further over his boots than the regulation six inches. He also wore his beret in Airborne style. All Airborne berets are folded over your right eye, almost completely covering it. He began inquiring about me as soon as he helped me get my bags. What surprised me was that he was picking me up, in his car, and helping me get my bags. I had never had a sergeant pick up anything for me, let alone help me. What the fuck was going on? Was this…respect?

The conversation quickly went from "Where did you go to basic training?" to "Why are you arriving from Korea as an E1?" I fumbled over my first few words before realizing I was slipping back into old habits, trying to excuse my behavior. Before I finished, I just spat out the truth. It felt like a hard sneeze. It hurts when you do it, but your nose thanks you. I told him I had fucked up and been punished for violating pass restrictions. His next response surprised me.

"Oh really? You'll fit right in then. You're coming to our platoon. We're called the Troublemakers, because everyone got an Article 15. We're misfits, fuckups, ya know? Of course we got a couple new privates, and they don't have any yet, but they'll have 'em soon enough."

I wasn't going to be judged for my past sins. This was looking better and better.

"Just keep your head down, you'll get your rank back, and everything will be good."

"Oh, roger that sergeant, I'm not going to make any more trouble."

"Oh, don't be a bitch, we don't need any kiss-ass motherfuckers!"

"No, sergeant, I just meant, fuck! I guess I'll go kill a hooker this weekend then."

SSG Fili laughed heartily.

"You'll be fine. Now, when we get to the company, I gotta be an asshole, but don't hold it against me."

"I can't promise, sergeant."

He laughed again.

We got to the company, and he began grabbing paperwork. He told me to sit down while he went to speak to the 1SG. Apparently I had been promoted en route and was now E2. They gave me my mosquito wings and an in-processing packet. I was then told to go to each place on that packet and, once I finished, I'd be good to join the company. This was something I was very unfamiliar with. I would have autonomy for at least a week. I was supposed to be responsible enough to go around this base, keep a few appointments, then hit up each location on this form on my own. What sort of world had I walked into? I kept to the schedule and, before I knew it, I was in-processed.

SSG Fili drove me to CIF (Central Issue Facility) to pick up my gear, and a few other places, and quickly I figured out the lay of the land. We had 4th ID on one end of the base, and 1st Cavalry Division on our side. We were 3rd Heavy Brigade Combat Team, Bravo Company in 1/12 Battalion. This meant we were closer to the 4th ID side of Hood, since it was split. The large in-processing station was III Corps of the Army and ended up churning out people like General Petraeus and General Odierno. We were close enough to the center of the base that most things were easily accessible. But my favorite location was right across from the barracks, the Class Six. These are liquor stores in the Army. They stay open 24/7 most of the time and sell alcohol at discount rates due to the tax easement they receive.

After finishing in-processing, I came back to the company and was introduced to the 1SG and Commander, 1SG Captain, and Captain Chapman. There was something off about them, but I didn't quite pick it out. From my understanding, this company was on the rough side of things. They weren't as clean-cut as the 1SG and CO teams I had seen in Korea or at Benning. There was an air about them of working dirty, almost like we were in the Mob. I remembered something my first squad leader, SSG Dillbeck, had said: "If you're seeing the 1SG, you're there to get an award or you're in trouble. You'll know beforehand you're getting an award, so most likely it's going to be trouble, so stay the fuck away from them." I made sure to heed his advice, and I interacted with these two only if it was a requirement. Already the battalion and company seemed fucked up. Our physician's assistant for sick call would constantly tell everyone they were lying and refuse to provide care. By the end of our deployment, he was kicked out of the Army for malpractice. Our headquarters company constantly lost leave forms and other paperwork and was closed most of the time. We didn't conduct sergeant's training on Tuesdays as usual, and most of our vehicles and equipment were broken.

I was sent to meet my platoon sergeant, squad leader, team leader and so on. This platoon was already set in a weird way and over the next year would get reorganized multiple times, but as it stood we had SFC Reynolds as our PSG. He was a black man with a mustache like Danny Glover's. He had just come back from The Trail, as we called it, slang for being a drill sergeant in basic training. He had been stationed with a POG training unit in some other base in Georgia and would regale us with tales of the complete pussies he had to deal with on a regular basis, and his inability to beat the shit out of those privates because they weren't infantry.

This platoon was split in two, because we were a mechanized unit. The platoon consisted of (to the best of my memory):

SFC Reynolds
Doc McCollough
SSG Embry
SGT Green
SPC Wall
Daugherty and Seiger (1st Bradley Team)
SGT Shaffer
SGT Sisson, Berg, and Johnson (2nd Bradley Team)
SSG Fili
SGT Rojas
SGT Sonkowski
SGT Raney
CPL Buck
CPL McGrath
Snead
Lopez
Myers
Barton
Birdsall
Hubbard

Brooks
Green
Simmons
Arenas
Miller
Martinez
Avila
Meadows (1st and 2nd Squad)
SSG Haas
SGT Crowley
SGT Zufall
SPC Nance
Davis
Stephens
Webb, and
myself (Weapons Squad).

We would get a few people right before deployment like Melsha — a 35-year-old PFC — but for the most part, these were the members of the Troublemakers, and regardless of position or rank, by the end of our deployment everyone had been a dismount or ridden a position in a Bradley crew.

SSG Embry oversaw the Bradley teams. There were two teams to a Bradley squad, and two squads. Each Bradley had a gunner, a driver, and a commander. On the flip side there was SSG Fili who led the other three squads, each with its own squad leader. Finally, SSG Haas was the WSL (Weapons Squad Leader, pronounced weasel). SSG Embry was a tall blond man with narrow eyes. He spoke as educated as an officer, but with the same hate as a drill sergeant. There was an air of affluence, and you could tell he thought he was better than you, no matter who you were. Whereas SSG Haas could not be any more different. He was slightly taller than SSG Embry, but very dark. His head was huge and his shoulders were broad. The man could fit at least half a can of Copenhagen in his lip from one pinch. He was massive and always looked pissed off.

He was from central Texas, and his accent came straight from a ranch. There was no affluence to his character, but there was a seriousness. Eventually he would be my squad leader, but for the time being I was placed under SSG Fili and kept off the Bradley teams. I would grow to appreciate this, because the Bradleys were hot death traps. These tin cans could barely fit five people inside, yet they always crammed a whole squad in the back. You were packed like sardines, and the ramp would lower so slowly you could time the Darth Vader theme song to it.

I made friends with most of the people in my platoon and hung out mostly with a new friend I still call my best friend today, Joe Davis. The squad consisted of him and me, Stephens, Snead, Lopez, and Nance, our Team Leader. Nance was a tall, dark-haired, hairy man from Dallas and always had a five-o'clock shadow. He even once had to shave in front of our Platoon sergeant to prove that he had shaved. Stephens was our goofball, Snead was our resident fuck boy, and Lopez was a chill Hispanic guy who just liked to drink and party. PVT Webb had been moved to our platoon because of a drug charge, and we joked that he was as hairy as Robin Williams. Then there was Sieger, who was in the Bradley squad but hung out with us regularly. He was our favorite to pick on. He was Mormon and hated it, so he would try any and all sorts of ill shit that had been denied to him by his crazy religion. There is almost a cultural requirement in the infantry to do your best to force pious or overtly religious people from grace and sink them to your level. The infantry isn't the company of philosophers or chaplains, it's the company of thieves, liars, brutes, and criminals. However, these are the only men who will willingly lay their life on the line and kill for you. It's almost a survival reaction to force anyone who would question their base instincts on something as useless as morality or piety. Seiger's sister was extremely hot and fucked a lot of the guys in the barracks. Last, Joe and I were very close in character. He was the brother I never had. We both enjoyed drinking, playing pool, making jokes, and most of all making asses of ourselves for nothing. The entire point of our humor was to out-psycho one another, just to see if there was a glimpse of humanity inside. I think we both found that to be lacking, and it was why we got along well.

Being part of the Troublemakers meant you resented authority. We always had bullshit to say, we always cheered for the underdog. When the largest soldier in our unit, Casey Hellenberg, competed at the Brigade Combatives Tournament, he went through every candidate in his weight class with ease. There was a tanker captain, though, who would fight others and act like a douche after winning. He was all ego, and when Hellenberg entered the final bout with him we began our animalistic cries. We shouted his name and stomped and cheered. The feeling wasn't that Hellenberg was our fighter, it was that we were excited to watch a private beat the shit out of a captain. Hellenberg had kept his fights interesting and struggled to challenge himself, and not just win, as a great showman would do. Yet once the bell was rung, within three minutes he had that captain on the floor in a rear naked choke and watched him pass out.

We found out once we arrived that we were getting a new pattern called the Army Combat Uniform instead of deserts, and they looked like shit. These "digital uniforms" were gray and felt like paper on your body. They didn't breathe well, you couldn't starch and press them, and you couldn't shine the new boots. This to me is the beginning of the end of military culture. Perception is reality, and these new uniforms made everyone look like shit. Everyone began wearing the uniform differently, because we weren't told the correct way to do it. The crotches were sewn so shitty that if you took a knee it would rip, so we all began taking knees in the field, and would have to go straight back to Rapid Fielding Initiative for reissue, because no one wears underwear in the field and our bare cocks would be exposed.

Then our company ended up getting put on funeral detail. Our job was to show up to old veterans' funerals in dress uniform, give their family the flag, and fire the twenty-one-gun salute. We hated this detail because you had to get in dress uniform and drive all over Texas. The perk was that we got extra pay for travel, were put up in hotels, and didn't have to do Physical Training in the morning. We went to a particular funeral, and like most times we arrived the night before and drove straight to the nearest bar to spend every dime of our extra pay. This time we got extremely drunk on fifty-cent beers, and one guy from a different platoon ended up taking a girl back to his room. We woke up in the morning, still hung over, and went to wake him up to get ready, and they were both stark naked in the room. The girl freaked out and ran away, and we arrived at the funeral ready to sweat in our dress uniforms in the Texas heat. As we began marching with the casket, we noticed something. The same girl from the room was there, in the front row. She was the granddaughter of the man in the casket, and the guy who had just spent all night having drunk sex with her was the soldier presenting the colors to the family.

During this detail Memorial Day came up, and we had to present the colors and conduct our duties for the holiday. We had to place flags, dog tags, boots, rifles, and helmets on the parade field for the fallen. It was here I got my first peek into how the heroes and veterans of this war were and are perceived. I saw and still see the disgusting behavior of my fellow countrymen running off to get deals, drink, shoot fireworks, and engage in any other action than reverence for the fallen soldiers that supposedly paid their blood for our freedom. Nothing was or is sacred next to the almighty dollar, and not one American gives two shits about dead soldiers. We sell the sanctity of the dead so you can get a three-day weekend and fifty percent off whatever junk you're trying to buy.

But the world still moves, and we would find any way to drink, and party. Even though we were underage, you couldn't walk down the barracks without running into alcohol. We had airsoft wars in the halls, brought women back for good times that ended up bad, and made a mess of everything we could. I would go downstairs and see one soldier parading a fat girl in a onesie with a sign that said "Shamoo Show," then walk down the other set of stairs to see Doc Brown in a banana suit outrunning the MPs to get them away from underage drinkers. We went to Austin to drink with college students, and I pissed on the floor of the ballroom in the Comedy Club. We drove on the beach extremely drunk in Galveston, and once I threw up outside of Snead's car when we tried to go to Padre Island and had to be dragged back to the barracks. The barracks parties themselves were out of control. People would hang on the balconies getting drunk until three or four in the morning. We got so drunk we forgot we used my Kevlar helmet to brace my door open and ended up pushing my door back with six or seven guys and crushing my helmet.

There were a few people that we were just terrible to as well. Birdsall was the thinnest man I have seen in my life and looked like Egon from *Ghostbusters*. You would be afraid even to shake the guy's hand because you might break it, and it didn't help that he was a bitch. He couldn't keep up in PT, was constantly crying while getting smoked, and to top it all off he was always wearing the weirdest shit. One day he wore a cowboy outfit, the next he would be dressed as a goth with a choker and makeup, the next day he was a punk. Life was a social jungle, and to survive you must be tough and, if the people around you sense weakness, they're going to jump all over you. Yet even though Birdsall was not the most attractive guy or the most macho, he pulled pussy like it came from a napkin holder at a fast-food restaurant. The women were always skanks, strippers, drug addicts, and worse, and after you slum that hard for that long eventually you get into the territory of finding out the girl is also a cheat. He ended up sleeping with another soldier's wife. This guy was deployed in Iraq at the time, which is already a huge taboo in the military, but to make the situation worse Birdsall knew he was downrange. As depraved as we were, just about everyone respected the unwritten rule that if you found out another infantryman's wife was going around turning herself out in the barracks, you were supposed to walk the fuck away. The man threatened

Birdsall by saying he swung both ways. He told him that he had a choice: When he came back he would either beat his ass or fuck it. Pichotta, the biggest lower enlisted soldier in our company, told Birdsall he would protect him. He said to tell the guy that he was Birdsall, and that if he came up to his third-floor room he would throw him off the balcony. If he managed to walk back up afterwards, then Pichotta would let him fuck his ass. We laughed, yet a few weeks later Birdsall took a fishing trip with that guy and didn't come back with any bruises.

To add insult to injury Barton ended up literally pissing on Birdsall when we were at the wash rack. He called him out because he married a stripper, popped hot for cocaine, then watched as she walked into the barracks and slept with anyone who would take her. He let anyone and everyone walk all over him, and we could only watch and speculate as to the moment the machine would force him to suicide. I felt sorry for him, but there's a survival mechanism that tells you you're safe if you aren't the focus of that kind of chaos. You don't want to draw fire, and at the end of the day he was living in a hell of his own making. His situation wasn't eased by the fact that he regularly hung out with the two weirdest people in the company, Stelly and Malitz. Stelly was another thin kid, but not as thin as Birdsall. Malitz was a forty-year-old man who was still a SPC and extremely creepy. He went to the strip club on a regular basis and dated strippers' mothers. He would pay for lap dances from the daughters but dated the mothers. He would also tell stories about liking his daughter sitting on his lap, and if you weren't too careful, he would pull out his TV and VCR and try to get you to watch tranny porn. We once pounced on Stelly because his grandfather had passed away. He started getting depressed, and we decided the thing to do was to call up Shamoo and have her fuck Stelly. We got him drunk, and she began talking to him. Meanwhile we ran out of alcohol and went to the Class Six. When we came back there was a car shaking in the parking lot, and we ran up to it to see Stelly's stick legs poking out around the fat folds of Shamoo. We began slapping the windows and howling until they got out of the car, and Shamoo thanked him for a good time.

We then deployed for JRTC (Joint Readiness Training Center) training to Fort Polk, Louisiana, the shithole of the Army, and that's coming from someone who has been to three different JRTCs. The air was thick and wet, the heat was out of control, the surrounding area looked like it was still in Civil War times. Houses were small wooden shacks that were falling apart, and the paint was peeling off the wood. The grass was waist high and the locals were not friendly, but we did have a random local donate some crawfish to us the first day we arrived. To make matters worse, one of the Bradley drivers had to go on emergency leave, so I was stuck with SSG Haas in a Bradley. I hated vehicles and swore I couldn't drive, but only because I feared being dispatched in a giant oven that would become paralyzed in a foot of mud. I had already seen the capability of these smaller-tracked vehicles in Korea, and it was pathetic. Every bump in the road, every obstacle was enough to break the track on these things, and you would spend hours resetting it. Meanwhile, if you were lucky enough to be a dismount, you might help, or more likely smoke a cigarette and get some air while the driver did all the work.

We ended up being crewed up with Daugherty, and he would help SSG Haas and me. He knew a lot about the vehicle, and SSG Haas used this time to get him to train up more on it so he could run it by himself. He ended up being a Bradley commander in Iraq. SSG Haas would act like he was quizzing him, but in reality he didn't know how to do shit on the Bradley because he had been light infantry his whole career. We sucked in that summer heat and spent hours laughing, telling stories, and waiting to do a lot of nothing. As SSG Haas explained, this was mostly for officers. They want to set up impossible standards and see how they react. An OC T (Observer Controller/Trainer) follows you around all day and tells you if you're reacting correctly. They can also "kill you" and set standards for missions. Part of this is wearing what's called MILES gear, fancy laser tag equipment that weighs twice as much, because this is the Army and everything needs to be heavy. We would beg the OC to kill us, and he would laugh. At the same time, though, these guys would be huge snitches. If you kept your hatch open to get some air, or weren't wearing your hard cover, they would go find the nearest NCO or officer to complain. I got caught a few times smoking a cigarette with my hatch open, and SSG Haas always kept cool about it. He would even look out for Daugherty and me so we could get some relief.

We played these asinine war games for thirty days, then they told us we were now acceptable to go to Iraq. We went back to the FOB (forward operating base), where the SGM was yelling at people for uniform violations even though we had just gotten back from the field. The thing that sticks out the most was the last day, when they gave some asshole POG an award for making her platoon donuts in the field. Of all the bullshit that occurred, the pain, the misery, the hate, this is what warranted an award. I rolled my eyes and gave the fuck up.

Upon redeployment to Fort Hood, the anxiety of the deployment began to loom over us. There was this feeling that it was time to let go of civilization. Soldiers who had deployed previously began to open up and impart their knowledge to us. We began being groomed for war, and every action or reaction was a test of our character. Our drinking increased, guys started getting married to anything with tits and legs, and those of us stuck in the barracks became insatiable children. Nance and I were drinking beers on the barracks balcony and started shooting bottle rockets into the rooms across the quad. One flew into a bathroom and exploded because apparently there was a meth lab in that room. We giggled in glee and ran, but were caught and sent to the company. Nance successfully negotiated our pardon because, as he put it, "We're goddamn heroes! We discovered and stopped a fucking meth lab!" It's times like these where the boredom of everyday living is the worst enemy you can face. We would create trouble just for the sake of doing so. Our leadership was expressing their frustration in acts of defiance such as pulling a lieutenant out of a Bradley and beating him in the face for aiming a turret at the tower. The lieutenant had high-ranking family members who were trying to set him up with deployment time for prestige and easy promotions. It seemed like we were invincible and this ride would never end, but everything came to a head one fateful evening when Snead, Lopez, and I decided to take a ride on I-35. We weren't drinking, surprisingly, but decided that our fun for the night would be to shoot cars with airsoft rifles on the freeway. Snead dared me to shoot at a car that had the driver's window down, and it caught the wind just right and the pellet struck the driver. We were pulled over and arrested but then let go. Snead was almost shot by the police for speaking to them aggressively and acting like a dick. Back at the base our chain of command had been alerted, and we were all in the doghouse. We were assigned to any and all ammo details, connex layouts, CQ and Staff Duty shifts, and any other form of labor that would keep us from going outside the immediate supervision of our team leaders. We didn't care; it was just more of the same bullshit, and after watching NCOs do heinous shit and not get in trouble, we victimized ourselves and blamed the system.

But in reality, no one can escape the wrath of the Army if it's too big of an incident. The toxic environment exploded when Malitz was supposedly caught downloading child pornography on his computer. 1SG Captain and Captain Chapman did something unimaginable: They called a company formation and put a hit out on Malitz. They told us that if it were their platoon, they would take him out back and take care of this problem, permanently. This was heard by his platoon, and everyone knew what was going to happen. Joe wanted in on the action, and I told him it would be better to stay out of it. Even though 1SG and CO were saying this, they would charge anyone involved. This wasn't going to end well. Sure enough, Joe couldn't sit by and went to Malitz's room with the rest of that platoon and beat Malitz within an inch of his life. I was next door and could hear the whole thing. I shut my door, closed my blinds, and smoked a cigarette and drank some beers.

The screaming, kicking, and scratching sounded like a horror movie, but there was no way I was going to do anything about it. I listened to the sounds like it was a concert, and once it was over I fell asleep. From what I heard they made Malitz crawl to the road until he was finally picked up by someone and sent to the hospital.

The outcry from this incident was immediate. Overnight we had NCOs and officers who began openly refusing to take direction from 1SG and CO. Others tried to fall in line and demanded the protection offered by our chain of command. IG, CID, and local police were always in our company, interviewing everyone. You couldn't walk two feet without seeing some form of law enforcement. Davis was caught up in this shit, and the entire weekend he was hauled back and forth from the police station to the barracks. Yet since we were so close to deploying to Iraq, this was quickly swept under the rug. We still completed our SRP (Soldier Readiness Processing) process and went about our business as usual.

The night before we left, everything came out. We began drinking the moment we were released and told our ETD the following day. You couldn't walk down the hall without being handed a drink. I saw Davis trying to hit on a girl and made my way around the units. The evening culminated with a company fight between A CO, Alpha Company, and us. The quad turned into a medieval battlefield, and the companies stood on the balconies screaming obscenities at each other. We taunted the other company, and they began making their way to the quad. Our platoon jumped up, followed the company downstairs, and began kicking, punching, and tackling anyone who wasn't from our company. I was struck in the face and began trading blows as well. No one was winning, but it wasn't about that. It was a spiritual event to brandish a war spirit. At one point I looked over and saw Davis smack dab in the middle of the clash. He was trying to protect the drink he had and the girl he was trying to hit on. As we looked at each other, a fist came out of nowhere and hit him in the face. He dropped his drink and became enraged. He swung for the first thing he could and knocked the girl out.

Another soldier began seizing next to the stairs, as Davis was trying to chase the girl down and apologize. The rest of us just laughed at the guy on the ground as the ambulance jumped the curb and nearly ran him over. We were laughing as the medics looked on in amazement and asked why we didn't help him. We just cackled back that our drunk asses would probably do more harm than good. Next the police began showing up and started grabbing people from the edges of the crowd. We began dispersing in such massive waves that they couldn't catch us. The sun rose and our broken, drunk, worthless asses stood in formation and dared the chain of command to say anything after the shit they had put us through. It was time for war.

IT STARTED WAY BACK IN HISTORY

COMMONLY KNOWN AS A 113

WHEN OUT OF The DESERT

CAME A BANG AND A FLASH!

IT WAS AN M2 BRADLEY KICKING SOME ASS!

CHAPTER 8

Step into the Sandbox

As I boarded the plane on the Fort Hood tarmac, I realized I had packed my brand-new helmet in the connex, but not my bags. I decided to do the shitbag thing and keep it to myself. I then sat next to Nance and began to sing the "Beer Run" song as long as I could, just because it annoyed him. Soon I passed out, then woke up, and we were touching down in Ireland. Ireland was the only country that wasn't welcoming to U.S. soldiers, and we were hustled into a large room and told we were not allowed to leave. I stood for what felt like hours, pushed up against the person next to me. We were like cattle in their pens waiting to be led to slaughter. Finally we left, and once again I passed out and woke up just in time to see us flying over Romania, which looked like a sea of connex boxes. Eventually I woke up again from my hangover and was told we were touching down in Kuwait.

We stayed a few days in some plywood huts in Camp Buehring. This base was just a jumping-off point for our next point before moving out to Iraq. We finally touched down in Ali Asilim, and soon a few familiar faces popped up: Tad, Mongol, Herd. And I couldn't help but think: Here is the meat grinder. Here we are ready to go forth and die. Or at least we should be ready. I was also worried that my level of shitbag and cowardice would be exposed. My new family would find out that all the outward display of insanity was a front, that just a few months prior I had been a failure and a pussy. But they didn't bring anything up, and I feel it might have been the commonality of the situation that restrained them. We were all going to a warzone, even Tad who had initially signed up only for two years. We were all about to face something awful and greater than ourselves, so why make it worse?

The short time we were there, we recalibrated our sights at the range and did some daily inventory and unpacking. I finally got my helmet, and no one was wise to the fact that I had left it in the connex. Score one for the shitbag. At the range, the guys who had shot expert and zeroed were allowed to just rezero their sights and leave. The rest of us stayed at the range all day in the Kuwaiti heat, just to shoot and continue to acclimate. At the chow hall, the range, the MWR, and even the gym I noticed something: There were a lot of civilians here. All these functions and more were staffed and operated by civilians. When we went to pick up our vehicles, civilians issued them to us. When we went to get the communications equipment, more civilians appeared. When we were issued our temporary barracks and sleeping site, civilians ran the booth. I couldn't find one thing on that base that wasn't staffed and mostly run by civilians. Even the night before we left, we had civilians coordinating flights out of Kuwait. The real joke was that, when we arrived at the range to conduct our reflexive fire training, the person leading the training was a Blackwater mercenary. Literal paid shills were certifying that we were ready to enter the sandbox; not an NCO, nor an officer. Not even a general was going to show up and ensure his troops were certified correctly, but a fucking overpaid hack was going to do the job the Army should be doing. I thought this was just a Kuwait thing. Kuwait had been used as a hub of operations for decades by that point, and there wasn't a war waging in Kuwait. I figured it would be different in Iraq.

Soon after arriving we lost our first sergeant, and he was replaced with 1SG Collela. This crazy guy spoke with a rasp and looked like his skin had been dried out in the sun for a few days. He understood what had happened and what the climate was surrounding our company. He promised not to get immediately into UCMJ

(Uniform Code of Military Justice), that we had more pressing matters. To his credit, the guys who were under investigation didn't start seeing military justice until almost mid-deployment. A few small fish like myself would get it faster. We were also introduced to our new officer, LT Ebarb. This guy was straight out of a magazine. He had gone to West Point, had family in government, had been raised in privilege. You could smell that he thought he was better than you. He got upset that people weren't giving him the respect he felt he deserved. He tried to get Arenas in trouble for petty comments during a morning PT run and constantly stated that he didn't feel respected. Arenas had two deployments under his belt and an injured knee from an IED (improvised explosive device). When Arenas couldn't keep up in a platoon run, LT Ebarb went to the rear and began hounding him. Arenas got upset and told him that he should see if he could run that fast after getting blown up. I think he eventually gave Arenas an Article 15. To say the least, I think Troublemaker platoon was not the right fit for a senator's son from West Point.

Any soldier makes sure, before leaving, to make one last visit to McDonald's. The greatest McDonald's is in Kuwait. It's the only American food within hundreds, if not thousands, of miles, and it's run out of a shack smaller than a taco truck. It's also the most expensive. In 2006 a Big Mac was six dollars, but you'll pay it. It's the only one there, and it might be the last one you ever have. Your last taste of America lies in the hands of some Arab guy who never even heard of McDonald's.

We also had to stop at FOB Anaconda in Baghdad. Here we would get our vehicles issued and drive out to our final FOB. Our final stop before our deployment began would be FOB Anaconda in Baghdad. I was awestruck that the place was huge. There was hotel-like accommodations, large chow halls, a fully stocked gym, a PX to rival Fort Hood. This was a place of deep history, the cradle of civilization, the first empires rose and fell here. There were rumors that biblical heroes might even be buried in this Fertile Crescent. I was able to see Saddam's palace and the victory arches and was struck by the lack of activity of the local population on the few trips we took outside the wire. Once again, everything was entirely staffed and run by civilians. You still had to drive around in military vehicles and carry your weapon, but you were safe. The occasional alarm was sounded for incoming fire, and initially we began scattering and heading toward bunkers and barriers. The guys who had been deployed previously laughed at us. They knew better. The weapons and abilities of most terrorists in Iraq could barely get an RPG to max out on its range, let alone fire a mortar or rocket that would land in the center of Anaconda. We had special anti-rocket systems on the perimeter, with gate and route security. The old guard stood their ground and laughed. Eventually the sound of the incoming alarm became normal, then an annoyance. The slightest breeze seemed to set the thing off, and people who hadn't been deployed previously or were POGs still ran and cried for cover every time. I understood quickly that in order to adapt to the new environment, you had to hold your fear at bay and keep walking. Grunts don't give two shits about mortars. If anything, our day just got more interesting if we did take fire. We weren't meant to be afraid; we were there to cause fear.

We boarded a Chinook at some dark hour of the morning as our Bradley crews mounted their vehicles and armed up to leave the wire. Being a private, I had no clue where we were going or what anything would be like when we arrived. I knew the base name was FOB Warhorse, and I knew we already had a mission, but I had no real clue about anything else. I didn't try to hide it or care, really. I was content just being pointed in the right direction and sent off to shoot as many people as possible. I was trash, and I knew my place. You could see out the back of the helicopter, and a few lights outlined buildings here and there. For the most part you had to use your NODs (night vision device) to see anything. The gunners were strapped in by a harness, and they all had different helmets and balaclavas. Each had some variant of venom teeth, or some monster like a werewolf or shark. To me they looked cool as fuck, because this wasn't regulation.

Suddenly there were a few plinks on the side of the ramp, and I realized we were being shot at. I was startled by the flash of flares shooting out the back of the chopper and watched as the gunner lit up roofs with his 240B. We began to cheer and laugh. More flares shot out the back, and the bird began rocking back and forth. One guy from a different platoon began throwing up, and we savagely laughed at him. After a few more rocks and flares, the bird took a nosedive. We hurtled straight toward the ground, and right before the moment of impact the bird pulled back up and softly landed. A few guys had joined the other guy in throwing up, and we stood up and made our way towards the tail end. We ran off the ramp at an angle towards the flightline, made our way behind some tall concrete barriers, entered a large trailer covered in green camo netting, and were introduced to an Army officer and his cadre of civilian contractors. They had an entire schedule lined out for us, and they briefed us on the base, where things were located, and where we would be staying.

We were housed in what they called Chus, hollowed-out connexes with AC units. There was camo netting along the rooftops and gravel between units. They smelled like shit. We were clear across the FOB from everything. The chow hall was two miles away, the MWR was three miles out, the PX was further. The only things we were near were a gate tower and the motor pool. That's what the Army does to the infantry: They keep them away from polite society. We are the rejected, the unseen savages that must be hidden. This isn't a phenomenon only on forward bases in a combat zone; it's true on bases in the U.S. and overseas. Dragon Hill was at the back forty in Camp Casey, in Uiejongbu we were located at the "Hindsight," at Fort Hood the infantry barracks were away from Headquarters and the medical areas. It's an unspoken reminder that you are not fit for public appearance, and I enjoyed it. If it was meant as disrespect in planning, then I liked it more.

Our platoon received another update and found out we would soon be split up. Half of us were going to be merged with a tank platoon. Our DCO was going to provide their last platoon, and we would give them our Heavy Weapons Squad. Before the official merger we would RIP (relief in place) in and take over for 4th ID. While getting settled in and assigned our Chus, Davis and I found out we were now being roomed with someone from a different platoon, Stelly. He was extremely dirty and stank. He wouldn't shower for days, wouldn't take his clothes to the wash, and his weapon was always jamming because he wouldn't clean it either. After a while we got tired of his shit, and we threw rocks and dirt in his cot and baby powder all over his clothes. When he came back in, he was angry and began throwing a fit. We pointed out that we had told him multiple times to clean his dirty ass up, and he wouldn't do it. We then brought out the fire extinguisher and had a fire extinguisher fight, and after that he went to our leadership and told on us. When we reported to SSG Haas about the issue, he looked angry at first but then busted out laughing. He told us to cut the shit out, and Stelly was moved to a different platoon. PFC Case then moved in with us, and he was twice as dirty. Murphy's Law.

Speaking of Murphy, here are Murphy's Ten Laws of Combat:

- Never share a foxhole with anyone braver than you are.
- No battle plan ever survives contact with the enemy.
- Friendly fire isn't.
- If the enemy is in range, so are you.
- Don't look conspicuous, it draws fire.
- If you can't remember, the claymore is pointed at you.
- The enemy diversion you have been ignoring will be the main attack.
- If your attack is going well, you have walked into an ambush.
- If you are short of everything but the enemy, you are in a combat zone.
- Tracer rounds work both ways.

I personally witnessed every law in this list. The only constant was that we were going to get screwed. Our platoon, in our company, in our battalion was the whipping boy for this deployment. A black cloud followed us around, and we intimately existed in the perpetual state of BOHICA: Bend over, here it comes again. This is not a gripe or a complaint, more a badge of honor. By the end of the tour our company and battalion had lost more soldiers than the others. In the Army they call you the tip of the spear, but by the end of our deployment we were the head of the hammer. We were dull, deadened, and assuaged. I'm not saying we lost soldiers comparable to the Somme or Gettysburg, but our losses were high, and it felt like we were specifically picked on or being punished. When you start hearing your squad leader, who has been in multiple deployments, tell you this is not normal, you tend to listen. Even when I speak to guys today about their deployments, I'll hear about a platoon losing one or two guys, or a company losing half a dozen guys. But having thirteen losses (not counting critically injured) in just your company, and your platoon takes the brunt of those casualties, tends to give you the idea that you're being targeted. That's why I want you to remember Murphy's Laws. Every time I start to get bleak, just remember the commandments we lived by. It will not bring you any hope or joy, but it might bring some understanding.

We reported to the company TOC to receive our first mission. SSG Haas was bringing me because I was supposed to sign my Article 15. For all the bullshit that had happened in the company it was Snead, Lopez, and me who were on the chopping block first. I would find out later that Snead's and Lopez's punishment would also be delayed until later in the deployment. The first blood sacrifice drew my name. I was going to feel the pain for the entire Troublemaker Platoon. Fuck it, bring it on. SSG Haas stressed that I must not complain or fight anything about the paperwork or anything, just sign it, get my punishment over with and move on, and everything would be okay. However, I felt gaslighted because in the rear 1SG Captain told me that if I fessed up to the crime my punishment would be reduced. He personally took us into the detective's office and parleyed for the charges to be dropped, and I quickly rolled over and put most of the blame on myself. I had shot the other guy anyways, so it was my fault. Yet here we were, new administrators, claiming the same shit, and I was going to get crucified. I guess since I was usually mouthy and spoke on a regular basis to legal about my rights and options, he thought I might be a tough cookie to crack. I just agreed and signed everything in front of 1SG Collela, and we made our way to the battalion so I could get read my Article 15 and then tried by my battalion commander. The Article 15 is an interesting event. It's more like going to traffic court, but often enough in the Army the offense is much greater. Your battalion sits and listens to your chain of command, from your team leader to your CO, on what happened, then they ask for your version, then they ask everyone in the room what should happen to you. Literally, your fate is decided by fewer than twelve people, who probably at this moment don't like you very much. I was yelled at, called a terrorist by the battalion, berated by the SGM. I sat and listened to my chain of command tell their version, and not one person stood up for me. The nicest recounting was by SSG Haas, who just recounted the facts and kept it at that.

When the battalion commander asked my take, I had only one thing left to say: "Sir and SGM, this was an accident. I was acting like an idiot, and I deserve to be punished. But this was an accident, and I didn't believe I could even do this. If you think I purposely shot that guy across four lanes of interstate traffic with an airsoft gun, you need to make me your personal sniper."

My fate was sealed. I was dropped in rank, my pay was cut, and I was sentenced to forty-five days of hard labor, extra duty. I left as I was dismissed, and SSG Haas met me outside. He was fuming.

"What the fuck was that, Russell? Are you fucking serious? You go in front of the battalion CO for an Article 15 and you ask to be his personal sniper? You are fucking insane."

"Well, I stopped listening after being called a terrorist and a monster for something so stupid."

"Go report to battalion for your extra duty, dumbass."

Soon enough I was placed under the battalion Ops sergeant major for extra duty, and it wasn't too bad. The first week I was just doing odd jobs like painting and filling sandbags. I also noticed something that week: There was another soldier on extra duty, a girl, and not just any girl, but one I had had my eye on back at Fort Hood. We called her Pancake Chick because she served the pancakes during breakfast. I would joke with her while getting my food in line. Everyone knew I liked her, but I would never shoot my shot, so to say. I also noticed she did not do any of the detail work I did, and the Ops sergeant major kept her close. I didn't think anything of it, because I'd never seen a girl on extra duty, and I thought he was just trying to keep her away from the men on extra duty. Had I been smarter, I would have realized that I was the only man on extra duty.

Our first mission dropped, and I was stunned to find out that I would have to keep doing extra duty while going out on missions daily. I laughed it off and headed back to the TOC to find out what was going on next. Our squad was told to head to the battalion TOC for our brief and found out 4th ID was giving us our left seat/right seat rides that week, and we would be the squad going out with 1SG Colella. We were to meet at the battalion TOC in the morning after chow and were going to use 4th ID's vehicles and weapons. We were to bring our own personal weapons, but all the radios, gear, and everything else was theirs.

We ate chow, slept, woke up early, and arrived at the TOC to see 4th ID pull up with a vehicle that had a shattered window and blood all over the seat. We were told the soldier in that seat had been shot just a few hours earlier. To make matters worse, that seat was my seat. I was the lowest-ranking soldier, and no one else would take it. And the conversation carried on without any regard to the fact that their fellow soldier had just died a few hours earlier, in that seat. Another day at the office. What the fuck was I getting into? We were going to do a cordon and search of the city center in Baqubah. We had the squad leader and gunner from 4th ID going with. The gunner was in our HUMVEE, and the squad leader was in the front vehicle with 1SG Colella. We went to the gate, checked in on the radio, and at that time SSG Haas told us to change our status to Amber. I did not know what he meant, but everyone began loading their weapons and charging them. I had the SAW, so I started to charge my weapon and place the belt on top. Nance stopped me.

"You realize if we hit a fucking bump that bolt is only held back by a small sear pin, right? Keep your bolt forward and place the rounds on the feed tray. When anything kicks off, charge the motherfucker and go from there."

I laughed nervously and did as he had told me. The gunner for 4th ID ensured he would take care of me and show me the ropes. The initial part of the mission was relatively uneventful. They showed us routes that were safe, meaning not littered with IEDs. Suddenly the conversation turned to the real goal of our mission, which was to capture al-Zarqawi. This would need to be something that later in my life I take with a great grain of salt. I remember riding with my 1SG and hearing Zarkari's name. I remember being debriefed that we usurped the leader of Al Qaida and that Baqubah was their capital. I also have spoken to my comrades in arms over the years, and have had this same thing parroted back at me. Yet it turns out Zarqawi was killed before we arrived. Whether or not this Mandela Effect was due to PTSD or PSYOPS spinning lies that confused all of us can never be known, but at the time we really thought we had just achieved something awesome. Davis had a worse challenge to overcome, because not only was this his first time driving in this war zone, it was also his first time ever driving. He was taking Driver's Ed in a combat zone, and getting ribbed by the whole truck while doing it. We pulled around to our portion of the cordon, which overlooked two roads and a palm grove. The gunner was showing me the turret and how to operate the crank, and SSG Haas was on the radio checking in with the other units.

While we were sitting around and shooting the shit, an explosion happened in the middle of the city. Without warning the palm grove behind us lit up from small arms fire, and the gunner above me began returning fire. The hot shells landed in my lap and all over my arms and face. As he was screaming in psychotic joy, he told me to lower my window and begin shooting my SAW. As I pulled the trigger for the first time in a combat zone, psychotic fervor poured out like water from a fountain. I smelled the carbon, heard the 240B firing above me, saw the brass flying in front of my face. I could not see anyone who was firing at us, but there were pings of bullets on the side of the vehicle. I looked out and saw members of the Iraqi Army shooting in the palm grove, in no real formation. Small mortars began to drop on their location and take them out one by one, and I witnessed a man take a direct hit from a mortar shell and get blown in half. There was a lot of chaos with SSG Haas trying to get Davis to drive and maneuver while a battle was raging, and before anyone found out anything it was all over. We did not really know shit about what had happened, other than that we had gotten to shoot at some trees, people were yelling, it was a real battle. I asked Nance if he saw the guy get blown in half, and he said yes. I saw the man's legs sticking out of the ground.

He then asked, "Is it real to you now?"

I said yes, but it really wasn't, it wasn't fucking real at all. We drove back and found out about the mission. We had fended off rebel fighters trying to protect Zarqawi. Something felt different, I had just been part of something, and it felt good. I might not have been the guy who cleared the house, bagged him, and sent him to be processed, but I was there to defend the mission that captured the leader of Al Qaeda. This was like being the guys who charged the Eagle's Nest to get to Hitler's bunker. But before anything could go to my head, I was getting yelled at by SSG Haas to get my ass to extra duty, and I was informed I could clean my SAW when I got back. The situation was unreal, and I could not wait to talk to the rest of the guys about it. I went to extra duty, worked a little more with Pancake

Chick, and tried to at least throw some looks at her so she might pick up on the fact that I liked her. I must have been coming off as a total retard, because she just put her head down and walked by faster.

After our RIP in we were paired with DCO and introduced to our new PSG. SFC Malo was a tanker who looked very unlike his mechanized brethren. Most tankers happen to be shorter, but SFC Malo was very tall and thin. He might have had their signature mustache, and what I referred to as Zelda boots because they looked like a medieval cosplayer costume, but he was fit and seemed more like a country boy who had grown up living in the woods. We were introduced to the platoon, and SFC Malo told us he would speak with us individually after our missions and his duties, but he would get to each of us. Present and accounted for we had:

SFC Malo
Doc McCullough
SSG Haas
SSG Davis
SSG Johnston
SGT Crowley
SGT Williams
SGT Green
CPL Barta
CPL Jackson
SPC Nance
SPC Burke
I want to say SPC Balzono
PFC Payeur
PFC Case
PFC Davis
PFC Webb
PFC Melsha
PVT Patrick
PVT Trussell
and me.

We were given a mission to patrol routes to the Buhritz IP (Iraqi Police) station and Embassy. Each day we would drive out and meet up with Troublemakers, then come back to refit and resupply. Eventually the tankers started to get to know us, and we were up to no good with the rest of them. We would play spades, smoke, watch TV in the Chus, and go to chow together. Things went through a bit of a haze, but I do remember small drive-by firefights on the routes we took. We would hear about attacks at the IP station and Embassy, and we would head out to respond, but by the time we arrived it was mostly over. I would come back and keep on with extra duty. It finally got out that I was into Pancake Chick, and it became a platoon issue, so the lower enlisted in our platoon hyped me up and convinced me it would be a good idea to talk to her. I tried one day, while we were painting a wall together, and we had what seemed like a normal conversation. I found out she was from Maine, she hated cooking, and she liked some of the same music I did. It was great, we had things in common, and I didn't feel like I had fucked it up. Well, when my shift was over, I was heading back to the Chus, and Doc McCollough appeared in a full sprint towards me. He told me to follow, so I did. We rounded the corner of the battalion TOC (where I just had come from), and he ran to a connex where I saw Pancake Girl splayed out on the ground, both wrists slit wide open, bleeding out. He started wrapping her up and working on her as I grabbed items from his bag.

Just then the guys in the platoon arrive and Doc McCollough looks back at me and yells, "You talk to Pancake Chick one time, and she tries to kill herself?"

Here was the girl of my dreams, and we were all laughing while she was bleeding to death. He pulled her from the container and took her to the aid station. The situation just got darker from there. Apparently, the Ops SGM had been drugging and raping her daily, and kept her on extra duty to ensure no one else got close to her. She was working full time in the TOC and being controlled by him. I don't want to think that me hitting on her caused her to snap. I later heard the SGM was demoted to SPC and sent to Leavenworth for his crimes. I never found out what happened to the girl.

After that a new NCO was put in charge of extra duty, and he was the kind of dickbag that believed everyone on extra duty needed to be punished harshly. Our new assignment became dumping refuse at the burn pit on the FOB and getting rid of the shit in the outhouses. Every day I would come back from mission and have to report to TOC and take a gator and drive to each unit to collect trash, then throw it in the burn pit. Burn pits are worse than you imagine. Not only is there trash, biological waste, rising smoke, and carcinogens, it's also a spot where you might literally get shot, not by enemy fire but by stray rounds, flares, and other explosives that lazy soldiers have thrown in there. I would hear the rounds pop off occasionally and see fires that never died down. It reminded me of the way Jesus described Gehenna as Hell in the Bible. The stench never ended, and your eyes watered as you got closer. I would fume inside, thinking I had just survived life outside the wire, only to catch a stray round in the fucking head from a burn pit bullet. To this day I hope that any soldier stupid or lazy enough to have left ordnance or rounds in any burn pit overseas dies a slow death from pancreatic cancer. The next part of my detail was to go to the post latrines and pour my weight's worth of JP8 (diesel gas) and light them on fire, then stir the shit and piss until it turned to charcoal. It was fucking gross, and I would come back soaked and smelling like shit. I was at least courteous enough to shower before coming back to the Chu, but of course some jackass would catch me halfway and yell at me for not wearing a KPOD (helmet) to the shower. When I was asked if I wanted to die if I were hit by a mortar, I would often say yes.

Meanwhile I could not keep it in my head that I was not the shitbag I thought I was. We had one sergeant who had deserted prior to deployment. He left on leave, and people joked that he joined the Mexican cartel, and other people had rumors going around that he was still at Fort Hood. Even still, I felt like I was a worse person or soldier than him. The details for route security started ramping up, and it seemed like Troublemaker and Blood Angels were the prime target for it all. I could not even remember the last day of my extra duty, but it became more apparent that we were a complete target for the battalion when a soldier from another platoon took early leave and deserted. He had been a hairdresser before the Army, and he was what they considered a perfect soldier. He looked like Captain America, was fit, got top scores on the PT test, was well spoken and very charismatic. Yet on a route clearance job he stepped out of his vehicle and an IED exploded. It was facing the wrong way, suggesting it had been hastily dug, and it exploded away from him. The story we got was that he froze. Other guys were reacting and moving around, but Captain America could not move. He staggered back into the vehicle, and after that he was not the same. He almost kissed death, and he was terrified. So he left and never came back. This guy was good-looking, had charisma and a future. I was an ugly fuckup who could not tell my secret crush I liked her, until her rapist caused her to attempt suicide. I did not have anything to live for anymore, and that is when I started to become a soldier.

We kept patrolling, which during the first month of the deployment was completely mechanized. We would take a few shots here and there and shoot back as we drove by, but we never fully engaged. I was told that a common tactic of the terrorists was to draw you in to some random guy shooting you, then start hitting you with mortars or IEDs when you hit a hard patch of dirt or a choke point. So we shot back at the men and kept driving. Until this point, I did not know if any shot I took killed another human being. I did not see the enemies at the palm grove, and we drove too fast to tell if I hit anything during our drive-by engagements. Now people were openly mocking the fact that my indiscretion back at Fort Hood had earned me the nickname "The I-35 Sniper" and I was now conducting drive-by shootings with my squad. The irony was that unless there were people I had killed on the first mission into Baqubah, the only thing I had "sniped" were clusters of camel spiders. They're not quite as big as the photoshopped images, but they were bigger than any spider I had ever seen. The biggest one was at least the size of a dinner plate. You could see them at night using the IR or the FLIR, and it would look like the entire top of a berm was moving, but it was those spiders. The prior deployed soldiers would tell stories of how they had Novocain in their fangs, and if you fell too deep into sleep while outside the wire, they would crawl on you and eat whole parts of your flesh before you woke up. Someone always knew "a guy" who had gotten his ear eaten off, or some shit like that. For fun we would go to the test fire pit and yell out "Test fire!" and start shooting the fuckers.

After about two months, we had been to the BIP on a regular basis and my squad was starting to learn the routes. Years later I cannot remember all the routes and sector names, but the one I remember is Route War Pig. Not only was it cool as fuck, it was the most dangerous road in Baqubah. The route was normally labeled black (meaning route clearance labeled the road as dangerous), and we took it as often as possible, because it was the quickest way to where we were headed. The first thirty days in country we had driven slowly, carefully surveying everything we came across, but after getting blown up a few times we became more reckless. Most IEDs are not what you see in the movies. Even the in brief you get when you first arrive in Iraq makes every IED seem twice the strength of the atom bomb dropped on Nagasaki. Most IEDs are placed hastily and without precision. The big bombs are set up by multiple people over a period. Someone is paid to inconspicuously dig a hole on the side of the road, then another guy is paid to buy fertilizer and paint, and yet another buys electric wire, then someone completely different is hired to pick up some or all of the pieces and deliver them to a bomb maker. Even still he will not put the bomb in the dirt; that's hired out as well. Finally, someone completely different buries the hole. The level of planning that goes into making a big IED and having it pay off is immense and requires a collection of data that takes a while. In the meantime, what happens more often is that terrorists pay off small-time criminals, farmers, and other people who are easy to prey on to dig a hole, then someone else goes and buries a quickly made IED. There is no training or qualification in this, and it's usually minutes before a convoy comes down the road. This hasty application usually results in a bomb not doing what it's intended to do.

I don't know how many bombs we would drive over and do a quick check of ourselves and our area, then keep rolling out, but it became normal enough that we stopped sweating the small stuff. If you saw large breaks in the road, or some serious construction, you would be tipped off, but a bunch of trash on the side of the road stopped scaring us. You began to collect your own set of data and learn the patterns of the land. You would see when people moved and when they didn't. You would see the same pile of rubble and stop focusing on it, or rather make sure it had not changed overnight. At this point you must face up to the fact that if an IED is going to go off, it is going to go off, and you need to rely on luck and your small amount of training to get through. The POGs that did leave the wire made a huge deal out of it, every time they did it. They acted like this was the last drive of their life, and even though the road had been cleared by route clearance, they were going to be crossing the no man's land in World War I. I laughed aloud every time I saw them setting up their anti-IED devices on the ends of their vehicles and bringing a priest to pray over them. I always thought: Give up, your God cannot help you now. You are in the land of the damned. You are stepping into Death's World, and it's time to put up or shut up. If you're too scared to play, then go the fuck home. I was ready to die. What was the worst that would happen, anyways?

Even though I was still heavily religious at the time, I didn't have any asinine ideas that Jesus was going to come down and save me from this. I had signed up for this, and no one was going to save me from myself. These asshats would come back and start some bullshit religious revival on the hood of their vehicles and thank

Jesus for keeping them safe. The Bible says those who live by the sword die by the sword. So it was my understanding that I had chosen how I was going to die the moment I signed up for this. The issue was the hypocrisy of the whole situation. Here were these assholes who rarely left the wire, who would scream at you on post for not wearing a PT belt or not wearing a cover from your Chu to the bathroom, but they would tell you what the safest routes were, or the best way to survive Iraq. They made up fake stories of their heroic deeds and worshiped the idea of fighting a war, that honestly they did not fight at all. The saying was that everyone wants to be Infantry, until it's time to be Infantry. Therefore, I kept wearing my flipflops and t-shirt back from the showers or to take a piss at midnight. If we were going to be attacked on the FOB, with anti-rocket weapons, gate guards, towers, and every defensive capability under the sun, then I should be able to make it to the john and back with no issue. But things in the Army are not supposed to make sense.

This brought out another issue. Melsha, the thirty-year-old private, who looked like he was at least fifty, would constantly wander off. We were told to always have a battle buddy, yet he couldn't understand this concept. He was too used to doing things by himself and taking care of himself. He would just wander off as soon as we were released, and not understand why everyone was in a full-blown panic when he came back. We would spend hours trying to find him, only for him to turn up at the Chus and be ready to go out on a mission. He would eat chow by himself, then arrive back at the same time we did, and we never knew exactly where he was going. He did not understand that the Army is not about autonomy. He may have been showing up, ready to go, but he was not with the team, and cutting yourself off from the team is dangerous. Denying the brotherhood is a mental and physical death sentence. Therefore, my first responsibility from SSG Haas became to not let Melsha out of my sight. I thought this was easy enough, until we would get back from the PX only to find out he had left for chow without telling anyone. Or he had left to get his paycheck from the finance booth and do his laundry, then when I left later and couldn't find him, he would turn up at these spots. This never changed throughout the deployment, and I felt like I was constantly being his nanny.

By this time we had gone through enough living with the DCO guys to earn our place. We would conduct patrols through town, dismount, and check local homes. The guys in DCO would function as extra riflemen at times or stay on the vehicles to provide cover while we patrolled neighborhoods and streets. We had cleared a few buildings together and taught them how to function as riflemen in a team. I had SSG Davis tell me specifically, "You're the expert, treat me as a Joe and tell me what to do."

I began telling a SSG which way to go when clearing a room and giving orders like a Team Leader. I could not believe someone of his rank was willing to listen to me, and the respect I got made me feel better about myself. When we arrived back at the FOB their guys would link up with us and we would clean up the vehicles, take them to the motor pool for maintenance, and make sure their tanks were in working order. We had not brought these bad boys out yet, but the plan was to start using them as part of our patrols and cordon missions. After a few training sessions on the tanks, they started taking us out as part of their tank crews and letting us drive, and even load the main gun. We would joke that we started a new MOS called Tanketry. At one point we were so packed with people and vehicles that we had an LMTV that needed to be guarded while on patrol. SGT Crowley was supposed to be TC (truck commander), and PVT Patrick was driving. Davis and I volunteered to be security in the back of the LMTV. This was extremely dangerous, because there's no armor in the back of an LMTV, and if we hit an IED we would be taking a direct hit. But we didn't give a shit, there wasn't enough room in the cab, and someone had to be ready to dismount and fuck shit up, and that someone was going to be Joe and me. We had to take the convoy through Trash Alley, where the trash burned day and night, and we knew terrorists were using the route to conceal their activities, because the heat of the fire caused the FLIR to become useless. It was just like the burn pits on base, but you were unable to use IR sights, which affected multiple weapons systems. I would put on my goggles and neck gaiter when we drove down this road to keep the ash and dirt out of my mouth.

One time we were ordered to perform a cordon of an area next to Trash Alley and SGT Green, our usual gunner, had left for emergency leave. I was asked if I would like to gun on the tank, to which I replied, "Does a bear shit in the woods?" We headed out. SFC Malo was the TC, Patrick was the loader, Burke was the driver. I thought it weird that, unlike the Bradley, the driver was not the private, but it was explained to me that the loader had the easiest job, and it was better to learn the tank in that way. The driver does not see what goes on in the turret, so if they have experience it makes it easier to maneuver and shoot the main gun. We drove through Trash Alley, and Burke had to lift the hatch to see. Afterwards we moved down a street to place ourselves out of the way for the construction of Hesco barriers, but also to give us visibility down streets and over the work. I was playing around with the FLIR (forward-looking infrared) with SFC Malo and realized we could still see the heat emanating off Trash Alley, but other than that, the FLIR was insane. You could see through walls, and miles away. I was told you could see the cherry from a cigarette butt at least ten miles out. This was different than any other sight I had used. I had used IR scopes for our machine guns, and the Bradleys had FLIRs, but they did not match the power and sophistication on the Abrams tank. This was such an understatement that even PSYOPS convinced locals that the FLIRS could see through anything but plastic. This resulted in idiots covering RPGs in plastic bags and hanging them out windows or around the corners of walls, thinking they could not be seen. Our tanks would see the person and just shoot the walls down on top of them, or we would flank the individual to get PID, then light them up with machine guns. We even had an instance where an individual had figured out the FLIR could watch them, and they jumped on top of a tank to try to disable the device. SSG Davis was in the tank and just spun the turret as fast as it could go, and when the man fell off the turret, he shot him with the coax gun. These guys lived by these sights, and if you had to use a reticle device or periscope, you might as well be guessing what you're shooting at.

Even the driver had access to the FLIR and used it to drive at night and in low visibility. The seat was comfortable, and I had more than enough leg room to stretch out. I was starting to think I had chosen the wrong MOS. SFC Malo showed me how to set up snap targets to move the gun faster and how to engage the main gun and coax, which was a 240C machine gun. Around this time, we got a call that the Hesco barriers could not be built any further, because it looked as though an IED or ordnance was sticking out of the ground. We checked it with the FLIR but could not get any positive ID on the material. The heat from Trash Alley was masking the rubble, making it impossible to see. We then called out the EOD (explosive ordnance disposal) team to arrive. Now, any soldier who has been deployed and spent time outside the wire can tell you that waiting on EOD is the worst thing in the world. It is one of the most time-consuming events you can be involved in. There are horror stories of people waiting twenty-four to forty-eight hours for EOD to show up. You go out on one mission that is supposed to be finished in a few hours, and it ends up taking days, and you don't even get time off, because you were supposed to get that during your wait for EOD, so now you must go back out again. I was caught up in these events more than once, so the thought of waiting for EOD was terrifying. You would get extremely bored and begin acting like idiots. SGT Crowley at one point was dared to smoke the tea from an MRE (meal ready to eat), and he did it. I was the designated gunner most times and was particularly salty one day, because no one would trade off with me on pulling security. So when SGT Crowley wanted a cigarette he offered, but I refused. My little rebellion caused him to start begging SSG Haas to leave the vehicle for a smoke. SSG Haas relented, but only if SGT Crowley would eat a cigarette, which he did. He vomited for at least fifteen minutes afterwards, but he got to leave the vehicle and smoke his cigarette. The event was so monumental that Nance and I told the story at his wedding ten years later.

You started talking about life and philosophy, telling stories, doing anything you could to break the boredom. Most guys would rack out and sleep in the HUMVEE, but I always kept watch. I never wanted to miss the chance to shoot someone, and I figured with my luck once I ducked down in the HUMVEE, some shit would pop off. None of that mattered at that time because what we forgot was that combat engineers were the mission. They got what they wanted when they wanted it. The infantry and tankers could take care of themselves, but not these poor precious POGs. So imagine my surprise when, within hours of the call, EOD showed up. We waited around laughing that the shit was a trash rubble that had dick to show for it, until we watched them deploy the robot. We nicknamed the robot Johnny 5, because of its likeness to the movie character. This thing rolled out of the back of a 113 and began heading to the rubble. No shit, there was a string of 155 mortar rounds strung together, ready to detonate and kill anyone working nearby. What made this instance different from the dozens of other times we had EOD take care of ordnance was that for this package there was a spotter, an Iraqi male watching everything going on. This is where I was tuned into the power of the FLIR. SFC Malo had me change my view and scan over, and I could see the man behind a concrete wall. He was bending over, then moving back and forth. I couldn't quite make out what he was doing, but then SFC Malo switched the sights to a daytime filter, and we could see the man swinging his arm. When we asked the team on site to see if they could get eyes on the man, they refused to get out of their vehicles, saying they feared the ordnance. While these bitches were busy shaking in their boots, we turned on the periscope just in time to catch a glimpse of the man trying to throw a rope around Johnny 5 like lassoing a bull in a rodeo. We began laughing. This idiot thought he was going to lasso Johnny 5 and drag a 500-pound robot out of there.

Either way, we could not let the enemy get hold of military property. SFC Malo told PVT Patrick to load a can round. He began to question it, until SFC Malo told him to shut the fuck up and do it. What I did not realize was that a can round is a 120mm shotgun shell with steel bearings the size of your fist. This round is not meant to be used on personnel, but for breach and equipment. Yet he had an idea of what he was doing, and we were going to fire. Patrick loaded, and we waited a while more. SFC Malo wanted to see if the guy was going to lasso Johnny 5, and to our surprise he finally did. I had my coordinates locked onto the guy the whole time, but the turret was facing the opposite direction. You could use the FLIR to set up targets and have a secondary sector with the main gun. So the whole time, the guy probably thought we couldn't see him. As soon as he lassoed our robot, he jumped up and down and did a little happy dance. SFC Malo yelled out, "Send it!!!" I hit the snap button and fired, and a loud metallic grinding squelched as the turret snapped to the target. The breach flew back and hit the back plate, and a crashing boom erupted as the round was shot downrange. The blast caused smoke to fill the air, and we could no longer see the area in front of us.

The EOD team thought the bomb had gone off and began screaming over the radio. The fun part was that it did, but that was because the giant can round set it off, as well as destroying the wall, the guy, and everything else. SFC Malo began sending up BDA to an officer and informing them of what had happened. I was shaking from excitement and laughing maniacally. I was now an infantryman with a tank kill. This was my first kill that I could see.

I HEAR THE CHOPPERS HOVERING
THEY'RE HOVERING OVERHEAD
THEY COME TO GET THE WOUNDED!

THEY COME TO GET THE DEAD...

CHAPTER 9

King of the Hill

This was glorious, and above reproach from any other infantryman! We began to return to base, and everything seemed normal. SFC Malo showed me how to clear the main gun and coax gun. We went through the process of getting ready to park and turn down for the night when, going through the gate, he told me to lower the gun. I had no clue where the breach came down, and rose, but while stretching out I saw a small hole at my foot that allowed me to stretch a bit further. This hole happened to be where the breach lowered into and could crush your foot, if you weren't careful. I began lowering the gun and felt a crushing pain on the top of my foot. It felt hot and intense. I immediately realized it was the gun, and I used the joystick to raise it as quickly as possible. It slammed against the ceiling, and SFC Malo began cussing and started to override the gun and lower it. I moved my foot out of the way just in time and was cursing in pain. He began laughing and called back, asking if the breach bit my foot. I nodded yes, and he said he would drive me to the clinic. They needed to see if the tank had broken any bones. As I writhed in pain, I could not understand how anything would not be broken. Yet at the clinic my foot was swollen to about twice its normal size, and they had to cut my boot off, but no bones were broken. I marveled and was given a few painkillers and a profile. I was to wear only soft shoes for a month. I could not believe it; I was going to be wearing tennis shoes on patrol. As I went back to the Chus I told everyone what had happened, and SSG Haas laughed at my misfortune. But I got a huge pat on the back for the tank kill. He told me to take the next day off and see how my foot was after some rest. SFC Malo then decided to create a platoon coin, a blood drop with wings, to signify the Blood Angels that was our namesake, and I still have that coin to this day.

Following these events, we would travel back and forth between the BIP and the Embassy on routine patrols, and often we would pass the Olympic Stadium near Baghdad. It was eerie to see this giant and modern sports complex empty and abandoned, starting to rust and showing a lack of care and upkeep. So we began stopping and conducting room clearing in the stadium. We found pieces of weapons, magazines, unspent rounds, and other materials associated with ammunition lying around. It was almost like we had just missed someone having a campout, who had scurried off and left their shit. This resulted in us starting to patrol the stadium regularly. I'm sure there were night watches and patrols on the stadium, because sooner than later we were getting a full-on mission brief about the stadium and when we were going to raid the thing. I thought we would have a full company mission, but it was going to be a platoon from DCO, the rest of the Troublemakers and us. Originally, they wanted the other Troublemakers to clear the stadium while we pulled cordon, until Davis and I started throwing a fit. We openly protested that we had been the guys to find this place and knew every inch of that stadium. I could have sworn I saw 1SG smile, and it was switched up that we would be the door kickers for this operation. We were briefed on the time and date and sent to prepare and run drills. We set up our plans with SFC Malo, and LT Ebarb would poke his head in occasionally, but SSG Haas and SFC Malo would not put up with it. They told him to shut the fuck up until it was time to give the OPORD. So we continued to prep, and the day came to raid this thing fully. I might have misspoken about how we knew every inch of the stadium, because certain doors were locked or barricaded, but we had cleared this thing at least ten times. I could tell anyone planning this that no one was going to be there, because the loud noises from tanks and other vehicles would tip these guys off, and I would be right. What I did not know was that the drones were watching the area and confirming that this was a hot spot for weapons trading. They were watching people move in and out of the stadium with weapons and concluded that it was time to move. We saddled up and began moving out in Bradleys, tanks, and our HUMVEEs. I began telling Davis all this noise was going to alert the terrorists and there would not be a damn person there when we arrived, and sure enough, there wasn't.

What we did end up finding was the largest cache of weapons in Iraq. We caught these guys in the middle of transferring thousands of weapons, and they just ran for the hills before we got there. There were RPGs, AKs, grenades, ammunition, and mortar rounds. A few of the squad leaders began grabbing AKs and RPGs. We would take these weapons and put them in the back of our HUMVEEs for later use and as war prizes, but the rest of the weapons and ordnance were sent to CID for inspection and investigations. You might be wondering why we not only kept weapons but got away with it. That's easy to explain. We kept the weapons in case anyone messed up outside ROE. If some idiot gunned down a family of four in a car, we had an AK to say they pointed it at us. If we saw a guy in a crowd trying to shoot us, and we missed and shot someone else, and he got away, then we had evidence. It was and still is fucked up, but it's war. I am not going to make this pretty for anyone. All of us were quoted at least once saying, "I would rather be judged by twelve than carried by six." This war was not just, and neither were we. Most who had been there once or twice just wanted to come home, and others just wanted revenge for the brothers who had already died there. The new guys like me just wanted to kill. CID also did not give a rat's ass about weapons. They were worried about us taking money or war trophies. If you had a few AKs and bayonets, they didn't give a shit. This find was big, made the Army look good, and was able to tie in Iran to the equation. All the serials we were told came from Iran. We started hearing rumors about instability at the border, and how the call to jihad had gone out and Iranian freedom fighters were answering the call. I cannot say for certain that we stopped anything major, but had the Iranians had access to those weapons when they crossed the border, the winter and spring would have been a vastly different picture.

Initially we were given three interpreters. The one I remember most was Snake. He had the best grasp of English and had been fighting since he was a child. His story was that Saddam killed his family when he was younger, so he picked up a rifle and had been trying to kill Arabs ever since. He was Jewish, and this man would begin shooting before any of us. He did not follow ROE or niceties that coalition forces were required to maintain. If he had an issue with someone, he would take them behind a building or in a room and "talk to them," and come back out wiping blood off his hands. If anyone asked what happened, he would say, "Do not ask if you do not want to know." He was the best informant and would openly walk around with his face showing. In Iraq this was considered a death sentence, and we ended up losing at least two interpreters because their identities were leaked. Because of this the rest of the terps kept their faces covered and made great strides to keep their identities a secret, but not Snake. He wanted to be attacked just so he might get to kill one more Arab or Muslim on his way out.

Another interpreter looked like Voldemort. His skin was ashy, and he was bald. He would solidly translate what you needed, but he would not carry a weapon and was required to be protected at all costs. I can't remember the names of the other two. One willingly went to give information to Al Qaeda and ended up being hanged in the town square anyway. We came across his body while on patrol and cut him down. Even though he went to sell us out, he didn't deserve that. That is another reason all of us as Americans should be ashamed for abandoning these people in Iraq when we left. They put their lives and their families' lives at risk so that we could "sleep at night," and we spit in their face and refused to help them when we left. Every politician and military officer that made that decision and carried it out should be tried and executed. If you're reading this and you were one of those people, do not twist it in your head and make excuses: You deserve to die, because you failed the people that sacrificed the most for your lies.

This brings me to the Kurds, who were the best allies we had in Iraq. You could count on the Kurds for just about anything. They were the most civilized ethnic group in the region. Most were educated, spoke English, and were extremely hygienic. When the truck drivers arrived on the FOB, the Kurds would immediately begin cleaning themselves after they finished their post-trip inspection and check-in. The Arabs would begin cooking and making chai, and most of the time they would not wash at all. Every time we went to a Kurdish base they had a full security detail, allowing us to take off our gear and stop pulling security. They ensured we were safe in their neighborhoods and would assist us if any firefight broke out nearby. One time we had a few prisoners, and we decided to stop at a Kurdish base for a security halt and to gain some intel. We began to move the prisoners and try to figure out where we could house them for the moment, but the Kurds took our prisoners and put them in a tower secured by two gates. The back gate was closed off and had just enough room for someone to sit on the floor. The front cage was where the Kurds would put a dog. They would smack it around a bit and get it mad, then put it in the cage, and the dog would snarl and bark at the prisoner. Arabs hate dogs, and this was an effective way to ensure they would not try to escape.

As we were finished with our halt, we heard shooting at the gate. Everyone grabbed their weapon and ran towards the sound of fire. I did not even put my KPOD on, and I noticed two Kurdish soldiers walking back from the gate. They told us to calm down, not to worry, they had just killed Ali Baba, which was their word for a terrorist. I asked one of them how he knew the person was a terrorist, and he told me it was because the man was Arab. I watched two other Kurdish soldiers fling his body in the open courtyard, piss on it, and set it on fire. We began patrolling their neighborhoods as a courtesy, and I saw the effects on these people of Saddam's biological warfare. Anyone who wants to deny the gas attacks from Saddam can go to hell. I saw old men who had deformed, yellow eyes, people missing limbs, children with birth defects. We would stop, give them medical aid, candy, and food, and speak with the elders. I cannot stress enough that we left these people to die when we left Iraq, and we did nothing to ensure their protection or solidification of statehood. We turned around and officially labeled them as terrorists in Turkey, and to this day we're assisting the Turkish government in its effort to commit genocide against them. But I am sure you all sleep well at night, knowing this is what the government does for your freedom.

The next set of players in the coalition to get screwed by us were the Georgians. These guys were everywhere when we first arrived. They guarded the chow hall, PX, gates, towers, even checkpoints on the routes we set up. The Georgians were a different breed. Most were dark-haired, short, and extremely hairy and wore woodland Battle Dress Uniforms in the desert. These guys rocked huge caterpillar unibrows like it was the height of fashion. We teased them endlessly, and most people thought they were not very smart. I was on tower guard with a Georgian Army soldier. I tried speaking to him, but it was quite some time before he even acknowledged me. Most of the time he was on his radio, which looked like an old school World War II radio, speaking to other Georgians. However, at the end of the day we tried communicating. You quickly learn that soldiers are the same pretty much everywhere. He wanted to see my patches and uniform badges and trade for his. We also traded weapons for a brief time, and he kept insinuating that the AK was better than my SAW. I pulled it out and began shooting from the tower to show him mine was better than his. This caused the tower control to lose their shit and start screaming over the radio, asking where the incoming fire was originating. We kept quiet after that and stuck to trying to figure out how to communicate. Eventually I asked about his rig, which looked like some pre-Vietnam flak vest with a few sewn-on pouches. This poor guy pulled out his med kit, and it was just a Muslim wrap, a match, and a bullet. He signed to me and spoke in Russian, but I understood what he meant. You were supposed to empty the bullet, light the match, cauterize the wound, and then wrap it up. I gave him my med kit that day.

The Georgians had been running checkpoints and tower guard since we arrived. They were scattered around the Diyala province on various routes and at our FOB, but they abruptly left, and we had no clue as to what happened. We had heard rumors through PNN (Private News Network) that the Georgians were slipping at checkpoints and taking bribes from terrorists to look the other way in return for not being attacked. At this time Georgia was considered the most corrupt government in the world, and it was very curious how some IEDs were set off a few hundred meters from checkpoints without the Georgians noticing, since they would stay there for 24-hour shifts. The attacks subsided and the IEDs started showing up just outside of checkpoints. They were either not doing their job or getting paid off. In 2008 Russia attacked Georgia, and we left them completely defenseless. We threw a hissy fit and did end up flying the soldiers back to their country to defend it, but Dick Cheney did an interview where, asked why we would abandon an ally, he basically said "Fuck them." Ukrainians now would do well to remember how the United States treats its allies after they help our war efforts. The Georgians were replaced by the Ugandans, who looked pissed at all times. I later found out we lied to them and told them they would be conducting operations and fighting by our side, not pulling security for our shops and chow halls.

Part of being deployed is that you have prolonged periods of time where there is nothing going on, and you must find ways to entertain yourself. I've heard of guys who got degrees, or increased their max lift, or some other inane accomplishment while being deployed. We were not so lucky. We were in sector every single day. You had enough time to RTB, eat, shower, and prep for the next mission. This left little time to do much more than play cards, or if someone had a TV you could watch a movie. Yet the Troublemakers had a different idea of fun. We would have fire extinguisher fights, huff canned air, and play dangerous games like Five Finger Freddy. One day Davis and I were throwing knives, and it turned into a game. You would throw the knife close to your opponent's foot and try not to hit him. Then the other person put their foot next to the blade, then you tossed your knife until you could see how far apart you could get your opponent's legs. We were playing this game in our Chu when Doc McCollough walked in and saw me and Davis with our feet nearly spread apart in a full split. He yelled at Davis to drop the knife, and Davis acted like he was going to throw it. He yelled again, and Davis faked again. After the third time we gave in and told him he was a killjoy. This was the thrill we got to break the monotony and show that we did not care about our own safety or wellbeing. I had stopped caring whether I lived or died.

By late November we took control of the Burhitz embassy and IP station. We would run operations out of the embassy, and our leadership would go to the IP station for intel and other things. Locals staffed the police station, and most weren't even trained police. The intel we received was always wrong, and we would spend hours waiting for EOD to show up to some site that never had a bomb, only to go back and get blown up. One day I went inside the station and got some bread from the police. I saw how they made it and nearly hurled. They squeegeed dirty water off a table that looked like it had bacteria originating from the primordial soup. They baked the bread, but I refused to eat it, and even though I told everyone how it was made, they still ate it. The whole squad contracted dysentery a day later, and I started hounding Davis, saying he was going to die like assholes on the Oregon Trail. He would laugh and then shit himself.

There was an eerie feeling, like a fog, although you could still see a light down the street. Things had been quiet minus the pop shots and a few IEDs when, out of nowhere, we got a call on the radio that the embassy was under fire. We sped in our vehicles as quickly as possible but only caught the tail end of the firefight. Two trucks had rolled up and begun unloading RPGs and a PKM on the embassy. Our guys were using suppressive fire on the trucks, but once we arrived from the flank road we unloaded on the first truck with the M2.50. The other truck got away, but we decimated the first truck and its occupants. Everyone got back to the FOB, cheering their first kills and the thrill of battle. Everyone except myself. I was finishing my extra duty and did not gun that day. You might ask who got to break in my 240B and light up the enemy first. I must say, sadly, that it was Melsha. Of all people, that goofy old fuck got to shoot my 240B and fucking killed someone with it before I could. I shook his hand and put on the best smile I could, but I would be lying if I did not feel the sting of failure inside. My stupid actions back in the rear were biting me in the ass. After the random attack we started staying overnight at the Embassy and BIP. We ran a couple of patrols out of the Embassy, and the rest of the Troublemakers had taken over their portion of the building, since they had locked it down first. It was kind of weird to technically be on the outside of the platoon. I wasn't a part of mission planning and honestly didn't care. I'd listen to the briefs and pay attention to some minor bits, but at the end of the day I was going to go where SSG Haas told me to go, regardless of whatever plan was made. But to the best of my understanding 1SG Colella wanted to test out the capabilities of using our tankers as support by fire assets, while using infantry to maneuver during operations. Normally this is the job of the weapons team for your platoon, but with all the assets available, and still no CO in charge, we got to play how we wanted. We ran an operation to try and go after an HVT (high value target), and each soldier was given posters and a pack of playing cards with different terrorists' faces on them and a bounty from DoD for each. We all joked that if we caught the assholes on those cards, we would not get paid shit, but that was a good thing.

Mercenary activity was out of control in Iraq. Blackwater was free to act however it wanted and suffered no consequences. You might remember that, at the beginning of the war, Blackwater had a few of their operators hung from a bridge and their bodies burned in protest by the public. People were rightly appalled by the action, but they could not understand how vindicated those people felt unless you spent time outside the wire. Blackwater operators were and are mercenaries. They did not respect the laws of war, they did not respect the population, and they certainly did not respect the military. They would drive through the streets, shoot anything that moved, and constantly cause collateral damage. These guys did not care about values, lives, or anything other than money. They got to dress up like "cool guys" and use better equipment, and this was all happening regularly before the Nisour Square massacre. After going through the streets and beating civilians, destroying property, and even killing women and children, it's more of a wonder why there were not more contractors hanging off that bridge. I am not saying that I was not also morally bankrupt. We tossed candy at kids to make them fight, and I had a wrist rocket in my gunner's hatch to sling rocks back at kids when they threw rocks at us. We threw piss bottles at kids begging on the street and, if a civilian entered our walking patrol, we brutally removed them. Worse is that the contractors would not lift a finger to help soldiers on patrol and would act like complete dickbags back on the FOBs. They would cut in the chow hall line. Most soldiers had at least lied to themselves that they were sacrificing themselves for the greater good or to protect their homeland. It was not the truth, but at least that lie kept us from completely disrespecting any form of life that existed. SSG Haas would not protect me if I just walked down the street and shot a kid in the head, for instance. Blackwater did not care; they just wanted to ride the government tit and steal as much money from our society as possible.

We moved out to secure our HVT and walked through each neighborhood. We had been dropped off by Bradleys, then tanks were maneuvering along the cordon set up on the outside of Buhritz. Again I started joking about how we just gave our position away with the sounds of the vehicles and that the guy was probably miles away by now. People in the city were on cell phones and watching our every move. Back then they did not really have cameras on phones, but we were aware that they used them like radios to keep from getting caught. At one point SSG Fili listened to my griping and sent Snake to grab the phone of one man on the corner who was staring us down and talking, but he said it was not who we were looking for. The sun was beating down, and people began to drop from the heat. We had to call in medevacs for some guys to get rehydrated. Then we heard a Blackhawk helicopter overhead, flying around the city. Someone was sure this guy was still in the town. The radios started popping off, and I could hear the static and chatter. We ran down one street, almost completely dropping security, then down another. Finally, we turned back around and went the opposite direction. SSG Fili was in a fury screaming directions and yelling at the RTO (radio operator). As we made our way through the urban maze, it seemed like the guy must have escaped, because SSG Fili threw his hands in the air and screamed. He located a large white house with a decent-sized courtyard and told the whole platoon to meet there. SSG Haas showed up, and the other dismounts from second squad, and we began sitting around, drinking water and smoking cigarettes. Some guys were eating chow, others were trying to nap. I looked at all the NCOs gathered around the RTO, and they were arguing, and I decided to try a small nap too. This brought on the best sleep I've ever had in my life. There's this weird in-between area I felt and have heard others describe. You're aware of what's going on around you, but you're in a state of sleep. I could hear conversations and movement, but I was getting rest. Once you "wake up," you feel more rested than ever. It was right on the cusp of this sleep that I heard SSG Haas tell everyone we were going to move out.

I woke up, checked my SAW, and walked up to the gate. Davis and another private commented on how they did not even know I was asleep, because my eyes were wide open. I laughed it off and shrugged my shoulders. At this point I noticed the helicopter was still going and sounded like it was getting closer. Before I could ask anyone what was going on, we heard the greatest sound in the world. We called it the Fart of Death, and it is the sound of the mini gun. That "BRRRT" sound is letting loose rounds at 66,000 RPM and can be a sound of either terror or excitement. We began cheering at the Blackhawk and could hear it move in and out, but it was getting closer by the minute. SSG Fili shouted out that the chopper took out a car, and they had the guy, then immediately recanted and said there was another enemy on the move. We were about to stack up on the gate and move into the streets when the gate busted open. There stood our target with an AK-47, and I saw his face go from a concentrated look to utter fear. The entire platoon opened on the guy. SSG Haas screamed to cease fire, and there lay the second enemy on the ground. We moved back to the vehicles and the FOB for a job well done. It felt cheap that this was my first time to shoot an enemy with my SAW and see it, but I would take the wins where I could. I knew I had reacted, just as trained, and I would no longer question my ability to do my job when needed.

Over the next few days we moved back out to the Embassy, where our squad would leave out on patrol with the tankers, then come back to the Embassy to pull guard shifts. The hours were getting brutal, and people were getting caught falling asleep on guard. I understood how difficult it was, but at the same time, the possibility of kicking off a firefight was too exciting to let me just rest. I was sleeping on the roof right next to the OP I would pull guard at, and then felt a small drop of water hit my cheek. To my surprise, the sky opened. What I did not know was that this part of Iraq had a wet season, and it had just begun. It rained constantly for days, and the dirt outside turned to mud. The lack of plumbing had reduced the public to throwing their shit in the streets, and the gutters were filled with human feces. We would jump ditches full of black mud and knew that if we slipped, we would be falling into a giant toilet. Sometimes guys would slip, and their boot would fall in the shit, and those boots were stained from that point forward. I joked with Davis that if I slipped, I'd be burning my boots as soon as the patrol ended. I bought at least two pairs of boots once the rain started, because when it really flooded, you could see human shit floating in the streets.

As the rain hit my face, I found I could no longer sleep. I told the guy on guard that he could go ahead and take relief. I could not sleep, and there was no point in him straining to stay awake. I was wearing only my tan shirt, pants, and boots, but I did not care. My shift was not supposed to end until daybreak, so I kept my IBA and helmet underneath the base we had built for the gun. If I heard footsteps or the radio tell me someone was on their way, then I would gear up. I looked out at the palm grove and noticed a mist lying low to the ground. You could barely see anything, and when I put on the thermal sight I was completely blind. As the sun started to break I heard rustling on the roof and decided to throw on my vest and helmet. This was just in time, because the next thing I heard was the 240B start firing from the opposite OP. I could see the tracers hit the brush under the palm groves, then I began seeing flashes. That OP was getting attacked, and it was happening! These idiots thought they could take the Embassy again! I turned my gun and let loose rounds. I could not tell if the enemy knew there was a .50 cal just down the wall from the OP they were firing on, or if we were concealed that well, but it was about to be a bad day for some guys.

Later, I would find out that things like this were local nationals who had been paid by terrorist groups to attack us randomly. They would find pariahs in their culture and societies and lie to them by telling them that if they attacked the military their families would be cared for. Usually, these people were so down on their luck it was like winning the lottery for them. I was about to pay out to whoever won in the hundreds. I pressed the butterfly switch and walked my tracers in to where the other gun was firing. There had to have been some cover under the brush, because an RPG flew out and hit the wall and everyone woke up. Troublemakers started running around, trying to get geared up and dressed, until SSG Fili jumped out of his fart sack (sleeping bag), naked from the waist down. He pulled his KPOD on and grabbed his rifle and began shooting too.

At that point, all hell broke loose. Everyone joined in. No one was wearing their uniform, and no one was fully dressed. Simmons jumped up and just hung a sock on his dick. Every weapon system, every rifle, grenade launcher, even flare was aimed and fired at the underbrush. The crescendo of rounds peaked until the brush caught fire. Smoke began to rise, and 1SG Colella ran to the roof and told us to cease fire. Another machine gun burst let off, and we started again. 1SG Colella began screaming, asking what we were shooting at. SSG Fili said he did not know, he had not even gone to my OP or the other to find out, the excitement was too much. 1SG screamed again to cease fire, after another burst shot out from the far OP. No one walking on the roof could tell him what had happened, and he was now in a screaming match with SSG Fili. Another burst flew out and I followed, and we were all ordered to cease fire once again. Finally SSG Haas came to my OP, I told him what had occurred, and he ran over to SSG Fili and 1SG. From the roof they checked out the brush, which was now in a full roar of flames, and told everyone to get off the roof except for myself and the other OP. They sent a team down to the fire and eventually found a few bodies with AKs. Over the next few days we began fortifying defenses at the embassy, and sleeping on the rooftop was now banned.

All the while in Blood Angel platoon, we had been enjoying the freedom of not having any officer oversight. The tankers were halfway reporting to our platoon leader, while also reporting back to their company. Our platoon leader was more obsessed with what was going on at the embassy than anything else. I do not remember him on patrols or outside the wire, except the embassy. It was as if LT Ebarb were obsessed with that place, and I could see why. At the embassy he could rub elbows with important people, political leaders, higher-up officers, and this helped him feel important and pad his resume. What I did understand was that the laxity in military bearing was getting to him. You do not want to salute, stand at attention, or single out an officer in a combat zone. This could make him a target and compromise your security. Yet the local national officers were saluted in the building and treated how he was supposed to be treated. On the FOB he made it a point to demand his salutes and courtesies. I was also aware that he had been told on at least one occasion to stop that shit at the embassy. He just would not let it go. That day we were setting up OPs on top of the embassy, when LT Ebarb went to the top of the OP and demanded PFC Berg salute him. I was told Berg was anxious and did not want to, but Ebarb demanded it. He shyly stood up at the position of attention and saluted LT Ebarb. A shot rang out from a sniper rifle, and they were hit. As we began to react to fire, they were evacuated off the roof. Berg died that day, and LT Ebarb was paralyzed for life.

We got a few days back at the FOB at that point, and at the same time another kick in the nuts. Nance had been somewhat of a big brother figure to us grunts this whole time. He had been deployed before and was easy to get along with. Just like the rest of us, his life was a mess and he had marital problems. Back in the rear he and his wife had separated due to insane bullshit she had done while he was stationed at Fort Campbell. This included the normal crazy bitch moves like calling the police on him claiming domestic violence, fucking other guys in their home, and doing drugs. By now they were separated, but she would still call the old 1SG on occasion and make up stories of him going to Dallas to rape her or some other crazy shit. It usually happened right after he refused to give her money. This caused a bunch of stress on him, and one day back at Fort Hood he went mudding with a few guys and they came across an abandoned vehicle. He became overwhelmed with anger because he said the vehicle belonged to a guy who was a deserter on their last tour. He decided to light the truck on fire and shoot it a couple of times. This set off the fire department and a host of federal police and park rangers, and resulted in him getting arrested. To this day I don't know the full story. Some people I still speak with say he was covering for Snead, Davis, and Eric who lit the fire, and each has a different story.

At the end of the day no one was injured, and the only property damage was that shitty truck, yet Nance took the fall for the whole thing. There was physical evidence to suggest that others were involved, but Nance stood his ground and said it was all him. We had thought that, since he had been overseas with us long enough, this issue must have been put on hold, or let go, but it was just waiting to rear its ugly head. This also came right after a bunch of the guys, including Nance, were huffing air to get high, and I ratted them out. I have never snitched on anyone, but a week prior we had had a soldier in our battalion die from that, and they began huffing so much that Nance fell to the ground and was foaming spit from his mouth. I felt like this had gone too far and told SSG Haas. He berated the guys for it, and I was cast out for a bit. I understand the team feeling uneasy about me, but I did have the best intentions. I still apologize for it to this day, but in my mind the safety of my team was more important. What made the matter worse was that the FBI came calling for Nance while we were deployed. They had put a warrant out for his arrest and labeled him a fugitive. The company was ordered to send him back to the United States, and he was supposed to stand trial on federal arson charges. The timing made it seem as though my actions caused him to be recalled and brought up the issue again, when it was just our wonderful justice system behaving as dysfunctionally as it always has done. Nance left, and before he got on the bird, he said I had done the right thing.

Our platoon's time to secure the embassy was ending. After the multiple patrols, and intelligence gathered by various missions, we had identified another building that we wanted to secure and use as a point of operations in Baqubah. We found the building fortified and crawling with Al Qaeda members, and after our naked gunfight they overran the BIP and flew their black flag. With a win under their belt, they thought the embassy would be a good next step, and once again it was attacked. During this firefight, there were multiple moving pieces. I was placed in a tank with SFC Malo and put on cordon duty. I think SSG Haas knew I needed to be away from the team now, and he did not need to worry about what the team might do or not do since the whole Nance issue was not settled. During this battle SSG Fili maneuvered his squad and SGT Bowlby took over the next squad. They were pinned at one point, and while Snead was returning fire he was shot. Doc Maddi grabbed him, drug him behind a HUMVEE, and began moving towards the door to the driver. He tried yelling Big Country, to not back up, but he heard the opposite, and they almost backed up over Snead. He was evacuated quickly after that, and the embassy was held. By now we wanted to take the fight to the enemy. We were tired of getting attacked, and it was time for some rage to spill over. Things began to boil over in the company as well, since 1SG Colella started pushing for legal action against the guys who had beaten down Malice. SSG Lopez was brought back in off the wire, and more questions were floating around. Davis also ended up getting served his Article 15, and now we were both at the bottom. Life fucking sucked, and morale was dropping quickly.

We went back to the FOB, and to my luck my birthday landed on the day we were there. We had been outside the United States since mid-September, and this was our second month in country. I had heard rumors of people sneaking hooch into the Chus and even creating some. SGT Crowley told us stories of another soldier on his first deployment fermenting alcohol, but just like most things it was all rumor mills. I got to have a Salisbury steak and a near beer for my birthday. I was hoping we would be in sector and I'd get to shoot someone, but no such luck. What did happen is we got mortared. I was sitting there eating my mystery meat when the alarm began to ring. The robotic voice of "incoming" was echoing through the air, and people around us began screaming. My reaction wasn't fear or even anxiety, but pure anger. I couldn't stand to hear these pussies screaming and running. It's primal, like an animal in the wild that's injured signaling to predators that it's ready to be scavenged. Other animals want to leave the injured pack member for survival. Here, all the bullshit you hear from POGs was on fully hypocritical display. The saying that "Everyone is a soldier/shooter first" went right out the window. These people were afraid and running for their lives. The joke was that you weren't going to outrun or hide from a mortar. You could run to the overhead barriers, just for the round to shoot right in the opening and cremate everyone inside like an oven. I sat there laughing, and as some random POG across from me got up to run away, I yelled at him asking if he was going to finish his food. He looked back at me disturbingly, and I stuck my knife in his steak and put it on my plate. He hadn't touched it yet, and even if he had, I didn't give a fuck.

My squad sat there eating as some dick sergeant at the door of the chow hall started screaming at us to leave, desperately trying to get us to go to the bunker. SSG Haas and all of us gave him the bird and told him to fuck off. We were hungry, and if we were going to die, it might as well be while we were full. He ran out, and some other E7 walked in and "ordered" us to leave. We began slowly grabbing our weapons and gear. He screamed to leave it behind, and we called back that we couldn't leave our weapons like every fucking POG in that chow hall had done. What if we got attacked after this? He yelled one more time and closed the door. By the time we grabbed all our gear and were standing up, the alarm stopped. We heard the all clear sound, and slowly the Fobbits started making their way back into the chow hall. Outside a few buildings had taken some mortars, and the anti-rocket weapons had gotten the rest. The SFC and SSG that had yelled at us were now staring us down and started heading over. SSG Haas stood up with his M4 at the low ready, when it looked like a SGM or some other higher-up came over to the chow NCOs to congratulate them on evacuating the chow hall, so they forgot about us. We quickly put our trays away and got to-go plates for our leaders back in the Chus. I walked outside to see flares and other shit popping off. SSG Haas remarked that at least I got a show for my birthday, and we all laughed

Back at the Chu we were told to go to the common area where the first sergeant, XO (Executive Officer), and future CO would be. It was time for a briefing, and we were getting a new mission. All platoons would be involved, and this was our first company-wide mission. In order to establish a presence in Baqubah we were going to take back the BIP and make it our company outpost in the city. We would have to raid this thing at night, then hold the building. Again, there was all sorts of mission planning and prep going on, and for the most part I didn't pay attention to any of it. I just kept to the idea that SSG Haas would point me wherever I needed to shoot and go from there. We surrounded the BIP with Bradleys, tanks, and our HUMVEES. The black flag of Al Qaeda rippled in the sky, as though attached to a ship's mast on the high sea. I was in the HUMVEE with SSG Haas, and he was sending up that they were seeing movement on the FLIRs of the tanks and Bradleys. These guys were getting ready to go, and we were supposed to make sure no one entered or left the cordon. Throughout the night the Bradleys and tanks unleashed hell on the BIP and the building next door that we nicknamed White Castle. This pure white house was on stilts, and imaging from drones showed movement from people in the BIP to White Castle. It didn't matter, whatever firepower they had wasn't enough to counter the TOW missiles, the 25mm Cannon on the Brads, or the main gun on the tanks. We didn't want to destroy the BIP, so at first the Brads were shooting only until fighters started shooting back with RPGs. The tanks opened and decimated half the building, and at this point someone came on the radio saying a large group of men had made their way into White Castle, and other stragglers were headed our way.

At this point I saw men with AKs trying to jump the ditch at the sector I was watching and began to fire. Davis dropped his window and let out a 203 grenade, and that was the end of those guys. By this time the BIP was retaken, and there were a few left in White Castle. We were ordered to close in the cordon and our vehicle drove up the street, next to the BIP. I could see Bradleys shooting at the White Castle with their 25mm and some pop shots being fired back. 1SG was on the radio, and I heard him tell them to send the TOW. The rocket flew out in between a burst of 25mm rounds and exploded on the face of the building. We couldn't see through the dust, but on the radio someone screamed that it was still standing. At that point SFC Malo piped up and asked to hit it with a main gun round. We were told they were going to send a SABOT round. This is a high-velocity round meant for armor penetration, and it's heavy as fuck. I was excited to see what it could do, and SSG Haas must have agreed, because we moved where we could see the action. The round let loose but again had no impact on the building. By now we assumed the people in the building were dead, and that anything else would be cleanup. The rounds flew in succession until 1SG called a ceasefire. He was ready to move into the BIP and secure the building. He called for Troublemakers to move to his location, a building across from the BIP. SSG Haas told me to come with him to the roof, and Melsha to man the gun. I tried to protest. He pulled me out and let me know that he didn't want Melsha around the 1SG, because he was goofy as shit and needed clout to get us in the building to clear it. I agreed and we made our way to the roof.

When we got to the top, we could see both buildings and second platoon's PSG and the XO. SFC Hamilton, or Hambone as some called him, was a complete clown. He was a close compadre to our PSG, but the shit he did made no sense. He had also specifically found some way to get out of the Malitz debacle, even though he probably had a hand in telling SSG Lopez to go ahead with that beating. Even still we looked at him, and when the 1SG told us to come forward, he began 3-5 second rushing the roof. This is when you drop to the prone, flip to your side, and get up and run. The building was almost half a football field long, so he was able to maneuver at least three times before reaching the 1SG. SSG Haas and I just looked at each other and walked to the 1SG. There at the corner he was surrounded by the XO, the RTO, and a sniper, and safety was not a priority. We went to the front, where a few people on the radio were arguing over how to take down White Castle, and the XO was speaking some gibberish on the matter. They had blasted this building with everything, and it was still standing. They wanted it destroyed even if there weren't fighters in it anymore, as its presence obstructed our views to the east and south. I looked at 1SG and respectfully spoke up.

"Top, why don't we just shoot the stilts out from under the building?"

"We would, but the XO just called in a fucking JDAM to take it out."

"I bet you it's still there."

"If it is, we can shoot the stilts, but if that thing takes a JDAM, maybe we should hole up in that building instead?"

We laughed at the notion, and SSG Haas told our squad to pull back and the radio chatter made its way to the tanks and Brads. After they pulled back, we could hear jets in the sky, and the sound was getting closer. Normally the F22 Raptors would conduct strafing runs before we entered a city on a clearing mission. When these jets flew over, it was not just the scream of the engine that got you, but the vibrations of this giant flying metal missile creating a vibration in the air that went straight to your chest.

We watched as the white building in front of us lit up like the Fourth of July. A fireball rose from the roof and black smoke covered the area. The bird was on the radio requesting status, but we couldn't see. Before the smoke had cleared, the pilot was already cursing. The damn house was still there! In a fury the pilot requested a second run, and he flew back to drop the second bomb. We saw the same effect, and the dust gave way to a White Castle still standing in defiance. I began to laugh uncontrollably. This tiny little house was mocking every bit of firepower we threw at it. There was now a small crack going down the center, and SFC Malo was asking to hit it again. 1SG looked at me as I nearly rolled on the ground in laughter, then came back on the radio to tell a Bradley to move into place and aim for the stilts. Finally the castle fell with a huge crash, flinging smoke, dust, and debris from the center. Even then, it was only cracked in two. It looked like someone had taken an ax to the middle of the building, and a few bodies hung in the parts that had opened. 1SG then told SSG Haas to move our squad to the front of the BIP to clear the building. SSG Fili was on the ground and would maneuver his squad behind ours. I moved to the stairs and could hear 1SG yelling at SFC Hamilton to stop 3-5 second rushing, as it was making him nervous. I told Davis what was happening, and SSG Haas told Melsha and the other guys to meet us at the gate to the BIP. We didn't expect any resistance, as the FLIRs weren't confirming movement, and this would likely be more evidence collection and BDA than anything. I stood at the door of the gate, ready to breach and then pull rear security. I kicked it hard and it flew open, our two squads moved inside, and I fell back with them. SSG Fili's squad entered first, and there was an ominous silence about the room. One of the guys from Fili's group had a tactical flashlight on his M4 and began moving down the hallways to the first set of rooms. SSG Haas moved our squad to the other side as another team entered the BIP to control the main room and courtyard.

At this point I heard gunfire. I looked at SSG Haas, who ordered me to move forward, and I stacked and readied my SAW. I began to sweat, nervous about what would be on the other side. As I entered the room, I saw a man slumped over in a chair with an AK still in hand, and I shot him. He didn't move and was probably already dead. I moved to my corner and could see outside to the hallway and the next room. I then moved over the man, grabbed his rifle, and pushed him on the ground. I lowered my SAW as the team was ordered to move down the hallway, and we could hear more shooting as SSG Fili's squad moved upstairs. I moved to the bottom of the stairs until I was bumped by someone in the previous squad. Davis and I made our way upstairs and shouted that friendlies were moving up. SSG Fili told us to secure the room that his team had just finished securing, but at that point another shot rang out from the hallway, and a bit of concrete on the wall flew off close to my face. I lay flat on the ground and unleashed a full belt of ammo. Davis and the others stepped over me and moved to the room. As I let go of my trigger I saw another body poke out and began firing. There was screaming about crossfire, and I was told to stay put. The final squad moved down the hall, and I turned to cover their backs. They reported another two dead guys, then moved upstairs.

By the time I got to the stairs at the third floor, we were hearing an all clear. Most of the people shot in that building were probably already dead or dying, but we cleared the bodies and moved them into the courtyard. Part of our training was to shoot a body before moving past it, then once the EPW (enemy prisoner of war) team came through they would inspect the bodies and look for intel. Most of the time you had to make sure there were no grenades hidden under bodies, but we were kicking them aside and checking while clearing. You wouldn't want to enter a room that had a guy booby trapped, so we kicked them over as we walked by them. The EPW team started acting like this was training and rolling the bodies while pulling security, just to hear SSG Haas yell at them that they'd been secured and to start looking for intel. We gathered everything we could in the main room, and 1SG finally made his appearance. The next steps were going to be securing the building and setting up a base of operations. Before you could blink the coms were being set up, .50 cals were being mounted at windows, and the camo net was getting laid around OPs. I went outside with Davis, and we began smoking a cigarette. SSG Haas yelled at us because we took off our helmets and told us that the SGM was coming and to look alive. We grumbled but did it. The SGM never showed up.

SGM Harris was the battalion sergeant major, and most of us hated him. He was a POG, a combat engineer, and did not understand the Infantry or our culture. He was hyper focused on how we looked, not on the mission. He would berate soldiers for not having properly creased pants over their boots, even though you had been patrolling all day in the heat and he was in his vehicle with AC. He would yell at someone for not having eye pro on while in his vehicle without eye pro. Nothing mattered but appearance, and it showed in our performance. Our OPORDs were usually shitty and had little information to plan with. We had the most dysfunctional gear in the battalion, such as my gunner's hatch that had a hand crank, and no up armor, but he ensured before we left for the box that everyone's HUMVEE had all the correct and same stenciling on it. At one point I found some chalk and drew a "combat monkey" on my turret, a sweet-looking cartoon with a cigar and an M16. The SGM saw this on a drive-by and turned his entire convoy around just to make me remove it. He also saw that I had my sleeves rolled up and freaked out on me. Nearly every day it was ninety to a hundred degrees. During the winter months the rain cooled everything down to eighty during the day, but the humidity negated any cooling effect. I had sand in crevices that I didn't know existed, hadn't showered or brushed my teeth since our last mission, and had stained my boots with human shit, but you could be damn sure SGM Harris would blow a gasket if he saw me unshaven. To make matters worse, he feared going outside the wire. He didn't show up to comfort our company when Berg died, and he wouldn't show up when others did either. The only time he left was when it was beneficial to himself and made him look good. Even if he was an asshole, if he had made his way out to the COP to check on us during a firefight, or after a death, we might have had a bit of respect for him. Instead we resented him, and if at any point I had been on his security detail, I would have looked the other way if someone had tried to shoot him. That's the level of hate we had.

I bring this up because shortly after gaining control of the COP we expected him to show up, but he wouldn't. We spent the next month fighting back attacks on the BIP to control that ground, and not once did that asshole show up. We were having the time of our lives living like savages, but the man who was supposed to lead us wanted nothing to do with us. But you can be sure he got a Bronze Star with valor before the end of the deployment, and so did every E6 and above in the HHC company of our battalion. You know who didn't get awards? The lower enlisted guys who gave their lives. Maybe they got them down the line, or later as part of some campaign. Appearance is everything in the Army, and it appeared that he cared more about how you looked and his career than about how we fought. The reprieve from him was well needed, though. We were heading into dark times, and the last thing we needed was some asshole telling us we looked like shit as we fought for our lives. I went to sleep in the COP that night, and I woke up in a whole new world.

MY BUDDY'S IN A FOXHOLE
WITH A BULLET IN HIS HEAD
THE MEDIC SAID HE'S WOUNDED!
BUT I KNOW THAT HE'S DEAD

CHAPTER 10

If You're Looking for Trouble

SGT Crowley woke me up with coffee cooked in his canteen cup. I had fallen asleep to the rain and woken up in a pool of sweat on the cot we had placed in a dark room. He told me we were to pull security while the rest of the company went back to the FOB for supplies. They had a few of the tankers on some OPs, and SGT Crowley, Davis, and I were the only grunts left back at the COP for now. I was shocked. What kind of military operation leaves a team of guys back at the ground you just took? We needed to prepare for a counterattack, and a squad of personnel and a tank were guarding this building we had just fought for. Not to mention that I had just been sleeping in a room where only six hours prior we had killed people. I looked around at the rubble from our attacks and asked SGT Crowley where the shitter was. He pointed at the rubble and tossed me some napkins from his MRE.

If you know anything about MREs, you know the napkins are paper thin and don't do the trick. I tossed them back because I always carried a roll of toilet paper in my rucksack, and I carried my rucksack everywhere because I was the machine gunner. It made for a nice footrest while in the gunner's hatch, and you couldn't fit the extra rounds I needed to carry in an assault pack, plus the packing list. I always carried an extra 1,000 rounds of 5.56 belt for the SAW and an extra 1,000 rounds 7.62 belt for the 240B. The max load for a SAW gunner was something like 800 rounds, but the rate of fire for the weapon is 1,000 RPM. This means that if you get caught in a firefight or ambush you must dump your 800 rounds out the gate and suppress enemy fire. Then it's on the rest of the team or squad to resupply you. A full squad and team is supposed to have an AG and AB, but we didn't. I was my own AG and AB. We kept rounds in the vehicles, and everyone had at least another 200 rounds in their assault packs, but I didn't want to be caught off guard. Your machine gun teams aren't sprinters, and we moved slowly and methodically on patrols. What I'm getting at is that a minute isn't a long time, but it becomes a lifetime if you run out of ammunition. Since we already moved slower than the other squads, I loaded myself up with more weight and got ready to hunker down for a big fight.

The perk was that I also got to carry more personal items. I had POGIE bait (snacks), a few Gatorades, two cartons of cigarettes, one log of dip, at least one roll of toilet paper, and a drawing pad with pencils. I was the guy who could always come in clutch with just about anything. Smoking was also the culture. I could do anything in the Army with a pack of cigarettes. If you wanted to find out about a unit, go to the smoke pit. There you could find out who was in command, and who to talk to about supplies or work orders. Share a cigarette or two and you might even get someone to do a favor for you. I did this with the mechanics on a regular basis, and the result was my HUMVEE was the first in the company to get the new up-armor when it arrived. I've found that this networking ability doesn't transfer into civilian workplaces. Civilians don't like to sit around smoke pits and chat, and they certainly don't like angry men telling foul stories. Anyway, the point is that I had enough comforts to keep our team cozy while we waited for the company to come back.

It was almost half a day before the company returned and began setting up. We asked SGT Crowley if we could be relieved, and he said he would get back with us. Before I knew it night set in, then I was staring at a rising sun and the call to prayer was echoing through the town. This became normal, but we also tracked.

If the prayer sounded different, we would radio up to get a terp to translate. I asked again if we would get relief, and once more SGT Crowley said he would find out. Yet again the sun set and rose and the call to prayer rang out. It wasn't until that next morning when SSG Haas came by to check on us and found out we had been going forty-eight hours straight with no relief. SSG Haas freaked out and yelled at the other squad leaders, demanding to know why no one had traded out with us, and there were some grumblings, but mostly it was just let go. We were told to get rest, and I went back to the bombed-out dark room and went to sleep.

When I woke again the room was lit by a small light bulb in the corner, and chemlights were cracked in every other corner. This time was different, though. There was a silence about, and I couldn't see any light outside through the rubble. As I rubbed my eyes and felt the slime of sweat drip down my back from the cot, I could hear radio chatter in the hallways. Then I heard a loud bang, and dust fell from the ceiling. Everyone was grabbing gear and running in different directions, another crack and bang. We were being mortared. I found a gap in the rubble and looked out, but I couldn't see the explosions. I grabbed the chemlights and began getting my gear ready. I put on my NODs and grabbed my SAW. As I entered the hallways, I was blinded by the lights that had been put up and had to flick up my NODs. That's when I noticed an OP down the hallway light up. These fuckers must have been trying to take back the COP.

I found an open window and started shooting my SAW, leading the rounds into the other machine gun's line of fire. A few rounds flicked the concrete above and below me, and I ducked and ran towards the OP. That's when I noticed Myers at the gun unloading a full belt. A couple of others were standing around, and I screamed at them. They should have been clipping another belt to his current. Before anyone could argue, I grabbed an ammo can of .50 cal and placed it on the ground. I told Myers to stop for a second and stomped on the rounds to clip them together. Then I told him to keep going, and SGT Crowley came around the corner and ordered me to follow him. I walked over to him, and he said we were about to go out and meet these guys head on. I gave him a thumbs up. We ran down the hallways until we saw Davis and SSG Haas. They were getting kitted up, and SSG Haas was yelling at Davis to get the vehicle ready. Another NCO yelled that we were supposed to wait for the FISTER (soldier who calls in indirect artillery or air support) to return indirect fire. I turned and saw our FISTER on the radio. He was yelling, and I could see SSG Embry behind him, berating him for getting something wrong. I could only think, Yeah, that'll calm him down and make him speak correctly. I stepped towards the door and saw Webb walking up geared and ready to go too. We looked at Davis and laughed and I looked out the door. You couldn't see anything, not even the mortar explosions. There weren't huge fireballs and smoke like in the movies. The biggest danger is the shrapnel, and that's when SGT Crowley pulled me from the door and asked if I wanted to die. I told him I didn't care, and SSG Haas pipped back that he cared, because they would have to waste people on getting my dead ass out of there. I shut up and sat down. We waited until the gunfire and mortars stopped, then loaded up in the vehicle. At this point we were just EPW and BDA teams.

Mortars began to increase. We would hear the call to prayer and take a few rounds, then it would stop. Each day it was from somewhere different. We were watching our OPs at night and not seeing any unusual movement, or at least anything we could identify to shoot. I would watch the elders play baccarat and card games while drinking tea, and SGT Crowley would bring us coffee. The temperature was now dropping to 80 degrees or lower at night, and we were shivering because it was at least 100 degrees during the day. We would wear our winter gear at night, then take it off during the day. The extremes began to make people sick, but it didn't matter. We had only enough people to barely cover the shifts on the OPs.

All of this would come to a head when a final attack was launched in December. We had been fortifying the BIP and placing sandbags and other items for cover and concealment around the compound.

At one point we were ordered to break down a wall that had rebar and just told to use our E-tools. As we were in the middle of bitching about these, shots rang out from the OP above us. We started to move and shot in the direction of the OP, until we were ordered to go back to digging. We were confused and started back until more shots rang out. The wall we were trying to break down would allow soldiers to pass between the two sections of the BIP to man the company radio without going outside, but I didn't care about that. If anything, getting shot would just make things more interesting. It would be better than waking up to mortars every day. More shots rang out, and mortars started flying in the courtyard. We couldn't tell where anything was coming from and were told to keep digging again. After a while, the gunfire and mortars had not stopped, and we realized this was some shit. At this point a few other NCOs came down and called for specific people. They then traded them out for the OPs and brought the OP soldiers down to help dig. I guess they wanted their new hallway that badly, but I thought this was stupid. Then everything stopped. Over the next few days we were attacked continuously. Small bursts of fire, then mortars. Every morning at call to prayer a flurry of mortars rang out. Dust began to rain down from the ceiling, and we were screaming in psychotic laughter at how stupid it was that we were digging out this fucking wall in the middle of a firefight. SGT Fulcher came down and saw we were not wearing our gear and yelled at us. We flicked him off as he turned around and grabbed our gear. This might have been exciting for most people, but it went on for days. We were switched off the OP, into construction details (making sandbags, removing rubble, setting up OP mounts for guns), then moved to a six-hour rest period. If shooting started on your rest period, then you began shooting and didn't rest. This grind continued for days, until the FISTERs were able to ascertain where the mortars were originating from. They sent teams out to check the areas, and they found the tubes, but no people. The reports came back that the terrorists were filling the tubes with ice and placing a round on the top. Once the ice melted it would contact the primer and go off, but the fighter had been gone for a long time. They were able to set up multiple mortars this way and just leave them. Then it was Christmas Eve.

I woke on Christmas Day to the smell of carbon and sand flinging in my eyes. Barton was kneeling above me and shooting through a hole he found in the wall. I punched him in his thigh and I told him to find another wall.

"We're fucking getting attacked!" he screamed.

"We're always getting attacked, dumbass!"

"No! It's for real this time! Look outside, asshole!"

I looked outside to see men and vehicles driving by while shooting. Huge black flags with Arabic writing scrawled across them flew in the beds of the trucks. Another mortar hit the building, and I looked up to see a certain crack.

We had tracked this crack from the first time we entered the BIP. It was on a foundational pillar, and it grew bigger each time we were mortared. We made a game out of it: Every time we were attacked we would mark the crack. It looked like a toddler's growth chart. I saw the crack pass the last mark and ran towards my gear. I joked with Barton that this time the pillar was going to break and fall on us. He smiled and kept shooting.

SSG Haas came in and ordered me to mount up in one of the tanks. We were going after these guys in full force. I moved to the "backyard" and saw the tanks already going. SSG Haas and I jumped in, and I asked about Davis. He told me he was with SFC Malo and CPL Barta. We would be going to another side of the compound to fire off some rounds. I remember loading tank rounds, one after another, and hearing the blast above us from the mortars. The first time shook me, until I realized the top of the tank was armored enough that no mortar was going to penetrate it. We kept firing until SSG Haas yelled for a ceasefire, then exclaimed that we had friendly KIA. Each of us, before entering Iraq, had been given a battle roster number, and we had this number sewn into our Kevlar band, which helped keep your Kevlar cover on. We had an

SOP of keeping it to the left of our NODs mount so it was visible and everyone could identify the same place to find the number. The other side had your blood type sewn on it. I bring this up because at this moment battle roster numbers started flying over the net, and no one could confirm who was down. The chaos was turning into a whirlwind, which finally exploded when SSG Haas looked over and told me they had just sent up Davis's battle roster number. I lost my shit. I demanded to leave the tank. SSG Haas would not let me, and I began to protest. It was not until he screamed at me like a child that I sat down and shut up. I deserved it, and I knew I was throwing a tantrum, but this was fucked. I boiled inside and did not know what to make of what had just happened. SSG Haas told us to load the main gun and prepare for it to be fired. I loaded that round faster than any other I have ever picked up and was full ready to send another. That is when SSG Haas reported out that Davis was only hit, and it was actually CPL Barta that had been killed. I sighed in relief, but then sighed again because we had just lost our first Blood Angel.

On the other side of the COP, Davis had also been loading in SFC Malo's tank. SFC Malo was stuck inside the COP, and CPL Barta took command of that tank. They had driven out to where they could see a few trucks with mounted guns and immediately fired a main gun. It tore through the truck and exploded in the distance. At this point CPL Barta could see the guys on the enemy mortar teams, and Davis asked if he could shoot his 203 at them. CPL Barta said yes, and he began launching away. CPL Barta turned the tank gun, and they loaded another round. In between this he told Davis to launch a flare towards the enemy mortar team so other people could see where the indirect was coming from, and Davis readied the flare. CPL Barta then opened his hatch to get eyes on and confirm what he was seeing on the FLIR when a huge explosion was set off above the tank. Davis was thrown back in his seat, and he could see CPL Barta slumped over but still standing. He removed his CVC and crawled to CPL Barta. When he tugged on his torso, CPL Barta fell over in Davis's lap and his brain fell out of his head onto Davis. Davis began screaming and rolled him over. The gunner took control of the tank and began sending up information. The chaos was too much, and Davis stared at CPL Barta until the battle was over.

The entire company saw the enemy mortar team, and we turned our fire towards them. Within minutes the team had been hit by everything from TOW missiles to SABOT rounds. There was no BDA; the ground lay charred and barren. We moved back to the COP, and everyone dismounted. The NCOs jumped out of their tanks, and all medics came out to check out Davis and Barta. As they scooped Barta out of the tank his limp body fell to the deck, and they placed him in a black body bag. I watched until someone yelled at me to come help, then I helped them lower his body off the tank. Davis went back to the tank to help the crew clean the blood and brains out of the turret, but even then, weeks later they still found pieces of CPL Barta in that tank. We didn't even have time to think. We just moved back to our OPs, pulled security, and woke up the next day like nothing had happened. We looked at our tanker brethren and couldn't even muster words.

AROUND HER NECK SHE WORE A STUDDED COLLAR

SHE WORE IT IN THE SPRINGTIME IN THE MERRY MONTH OF MAY

AND ANYTIME YOU ASKED HER WHY SHE WORE IT?

SHE WORE IT FOR THAT SOLDIER WHO WAS INTO WHIPS AND CHAINS!

CHAPTER 11

I Saw a Pale Rider, and His Name Was Death

I stood at attention with an M4 in my hands. I had a magazine with three rounds of ammunition in it. The gymnasium was full of chatter. I stood there on rifle detail for CPL Barta. If I focused I could hear mumbling, some laughter, and other loud voices. To these people this was a fucking event. As soon as the chaplain walked up people quieted down, and music started. It sounded like a classic organ funeral sounding off, but it was just a CD recording from a boom box. The chaplain began speaking, then you could hear the sniffles and moans from outside. I knew our fellow Blood Angels were true in their mourning, but just like every time there were at least a hundred POGs who didn't even know you crying and screaming like they knew him personally. It made me sick to my stomach, but I just kept going. I was outside my body, and everything was just happening. We were ordered to give the twenty-one-gun salute, and to fire off our three rounds. This was surreal, and Davis was next to me looking the same way. I hadn't yet asked him how he felt, and I wasn't going to. If he needed to talk, we would all be there, but for now we had a mission, only instead of searching a town or clearing a building, our mission was playing the firing squad detail for your brother's funeral while people who hadn't known him pretended to be sad.

Afterwards the chaplain walked out and asked if we needed a prayer, and Davis and I declined. I wanted to tell him to fuck off and take his dog and pony show somewhere else, but I knew better.

I shut my mouth and walked off. We marched back to the motor pool to get on the HUMVEES and go back to the COP. At one point we heard other people talking about hitting the chow hall or PX before we left. I smoked my cigarette and stewed. This was a fucking reprieve from where we needed to be. We needed to be on patrol and storming every fucking building we could to kill any military-aged male we could find. I wanted to burn the whole fucking city to ashes, and I guess it's good I wasn't in charge of shit, otherwise I would have started committing war crimes.

Before we headed back, 1SG called us all over and began briefing us on things and intel going around. It was a ramp brief before we headed back to the COP, but he was throwing in extras for the time. We found out the border of Iran had lost control, and the call to jihad had gone out. Iranian freedom fighters were crossing the border to join the fight, and we would see an increase in operations soon. This other word was flying around: The Surge. Apparently we were expected to ramp up operations all over Iraq and would be getting replacements soon. At this point we had lost only two guys KIA, but our wounded was at least ten for the company. To our knowledge those guys weren't coming back, and this meant the tower guard shifts started getting longer, and the details less manageable. We couldn't even spare people sometimes to go back to the FOB for supplies. At one point we ran out of MREs, and 1SG didn't even notice until we told him. He called us liars, saying he checked the locker room every day, but when he went to verify, he saw nothing but egg omelet MRE. This was some dehydrated egg powder that was supposed to turn into an egg, but in no way did. He freaked out and did his own emergency run to the FOB to resupply, and we joked with him about it. You had to laugh. Things were getting dark, and if you didn't laugh you would be overtaken by despair.

Then came the need for lights and electricity at the BIP. We were still getting mortared at least once a week, and we would flick off the incoming explosions and scream. I began to scream out, "Rain god's a pussy! He won't make it rain today!"

It's almost as if thunder hit and mortars fell harder. A generator repair mechanic had come out with his team and was setting up the electrical work for our building. When the mortars rained down, he was caught outside and began screaming. He started barrel rolling over the generator and pulled wires and turned his tools. We kept wondering what the fuck he was doing, so a group of us gathered round to see the show. He began screaming to cover him while he moved, as the mortars mostly landed in the "backyard" of the compound. Even 1SG came out and watched for a second, and then we all laughed. He told him to get his stupid ass inside, and we would get to it when the mortars stopped. He looked at us like he had just done something amazing, and we commented on how we still didn't have lights. I'm sure he went back and told some story about how he acted like Rambo and got us up and running, but at this point mortars were a daily occurrence. I would be more scared if I didn't wake up and hear them.

We spent our small amount of downtime smoking and playing cards. At one point we were all pontificating on what things were bad luck. One guy said Charms (square life saver candy in your MRE), and I remembered back to LRS-D, where my team leader had seen me try to eat them and grabbed them from my hands. He threw them as far as he could, slapped me in the face and screamed that Charms were bad juju. Some other guys were saying the R word, rain. I remembered again back to LRS, where in the field they made us dance shirtless around a rock formation. We set up a circle like a turtle and gave offerings of cigarettes and dip and chanted "The rain god's a pussy." This was supposed to make it rain. I then brought up those yellow lighters that were supposedly bad luck, and Webb argued that it was white. We were laughing and calling each other names, and then I lit my cigarette with his yellow lighter, and he lit his with my white. A crack hit the wall next to my head and debris went flying. Another firefight started, and we began laughing maniacally.

During this fight we were shooting from every OP, and I was

moving up towards the OP I would be pulling my shift from soon. I could hear machine gun fire from the room and noticed there were at least ten people in the room. I yelled out asking what the fuck was going on, and apparently the OP was taking fire from a rooftop. Everyone and their mom was in the OP trying to fire back, and some asshat was on the gun and not hitting the guy. I walked to the gun and lit my cigarette, and the blast shield in front of my face cracked from incoming rounds. I smiled and set my SAW down and started shooting. Suddenly a crack rang through the room, and I heard someone screaming "Medic!!!" A guy from another platoon, Castillo, had been shot in the head. He lay on the ground for at least a minute, and not one person spoke. It was like the world just stopped. Yet he opened his eyes and sat up. His face was wild and full of fear. He timidly asked what had happened. That's when the two biggest retards in the entire platoon stood forward. SGT Fuller and PFC Myers jumped in and yelled at him that he'd just been shot in the fucking head, and then Myers laughed.

These two were the absolute biggest idiots to walk the Earth. For instance, Myers was on a vehicle patrol where he was supposed to pop a hand flare for the dismounts. You grab the tube, pull the cap off, place it on the bottom of the tube, and shoot it in the air. Instead, he slammed the damn thing against the side of the turret, shooting the flare into the HUMVEE. It looked like Tinkerbell was fighting inside the vehicle, and I had never laughed so hard in my life. But he was at least young, and from the country, so he had an excuse. But Fuller, that guy was a complete waste of sperm. He would constantly get on to people for shit he did himself, he was fat and could barely run, and he would try to speak like a black person even though he was clearly an asshole from some distant trailer park. He would eventually get investigated because on his mid-tour leave he shot himself in the leg while "cleaning" his pistol. We knew he was bitching out because things were getting heated, and everyone hated him.

Cass began screaming and Doc Madi came up to grab him and

take him to the aid station. I looked at both and screamed, "FUCKING RETARDS!" SGT Fuller got mad, but I just went to the OP and took over the gun. I let go of the belt and told the other guys to start linking other belts. We killed the guy, but the sniper was still out there. It felt fucked up to think, but now I was getting annoyed that we had one less guy to pull security and do details. I understand the next paragraph might be too much for people to take, but I swear I remember this occurring. I have heard I don't know how many tall tales from POGs who would lie about their escapades. They would tell stories about how they Ramboed their way through terrorists and cleared a whole town by themselves, even though they never left the wire. You would hear them talking at the smoke pit and could just hear the lies dripping from their stories. I couldn't stand it. If you wanted to be a grunt, you should have nutted up and joined the infantry. I knew better, knew these fucks just pulled tower guard, maybe saw the occasional mortar round. That being said, this is the absolute best way I remember this event. If I am wrong on details or confused, then so be it, but it was a wild story. Cass had been struck by a sniper, and we returned fire to defend the BIP. During other patrols this activity was increasing, and we were desperately searching for intelligence. At any time you could be the next victim, and I did not care. I hoped at most points of the day for some type of engagement to end the monotony. We returned to the COP from a patrol, and I saw 1SG and SSG Haas in the RTO room, and they began talking about the snipers. 1SG Colella spoke up and said they were two brothers from North Korea. I understand now. Then, the implication of North Korea operating in Iraq would have sent shock waves through the world. Whether or not this was PSYOPS intel meant to cause confusion or actual intel was beyond my rank or ability to discern, but 1SG was very clear we had obtained this from the Kurds and that the two individuals were operating in Baqubah. I was astounded and excited at the same time. The ability to possibly put a bullet in a North Korean sniper was enough to give me a bit of motivation throughout the day. I began keeping a closer eye on everyone during our cordons and listening to NCOs a little more carefully.

I remembered this even when later during a firefight SGT Fuller tried to launch an AT4 from a hallway out a window. An AT4 is a small launcher that sends out an explosive for anti-vehicle operations. They're lightweight and easy to use, but they have a significant back blast from the explosives and the way the tube is shaped. Fuller stood in the hallway with the AT4 on his shoulder and aimed it ready to fire. The back end of the tube was maybe a foot from the wall behind him. I smiled and watched, ready to see this jackass get launched out of the window in a giant fireball. Yet SFC Reynolds came around the corner and stopped him. He first screamed "Back blast area clear!" This signaled Fuller to realize he needed to check his back blast, but instead of realizing there was a wall there, he just repeated what SFC Reynolds said as he looked at the wall then looked forward, ready to fire. SFC Reynolds then screamed, "YOUR BACKBLAST IS NOT CLEAR!!!" Fuller realized what the issue was and threw the launcher out the window and started shooting at nothing with his M4. SFC Reynolds looked at me, and he had a look like he knew what I was doing and wasn't going to stop him. I just smiled and shrugged my shoulders. You win stupid prizes if you play stupid games, and he was shitting on the Monopoly board.

Finally we ended up with another lieutenant, another West Point graduate who was moving through the military with the intent of padding his resume. At this point the only stable officers in our ranks were 2nd Platoon's LT and our XO. Our communications with 3rd Platoon were limited, and I never saw them unless we did joint operations, and 2nd would trade with us at the embassy, so we at least saw them on a regular basis. We knew they both had a PL, and the first one to arrive was supposed to be dibs for us. Yet this guy showed up with some goofy-looking glasses and was out the door before we even knew his name. I heard he was in a HUMVEE during an IED blast and then decided to move back to one of the S Shops. We would have to wait a bit longer for our next LT to come to the rescue.

During this time, we would run at least one mounted patrol per day and conduct what were referred to as Snap TCPs, traffic control points. We would drive around until told to stop, then set up concertina wire in the road and begin checking vehicles. There was an increase in IEDs on the road, and POGs were tired of getting blown up. We had signs with Arabic writing on them, with stop signs, flares, and other means to direct traffic. The signs said to stay back at least a hundred meters, and we set this sign up at least a hundred meters away from our vehicles. This meant the traffic was supposed to be at least two hundred meters away. Then we would move the concertina wire to funnel traffic one way or another. This was intended to try and catch cars transporting IED materials or weapons. Each person in Iraq was allowed to carry one rifle for protection, but the real things we were checking for were RPGs, silver paint, tarps, and fertilizer. We had been conducting TCPs all day when 2nd PLT showed up and assisted us with a small cache of weapons we found. As we were moving the weapons into another vehicle and taking pictures of everything a complete asshole decided to herd his sheep into our TCP. They became entangled in the wire and were bleating and causing a mess. We were holding up traffic more than usual, and the shepherd was acting amazed like our TCP just appeared in the middle of his herd. I began to scream at him to move his animals, and he kept replying "Mista why? Me love America. Why mista?" I couldn't take it anymore and threatened to shoot a sheep if he didn't get moving. The whole time Snake was relaying what I was saying to this man, and then turning back to tell us he was an idiot and trying to blame us for the problem. During this fight Pichotta came over and saw one of the sheep that was too tangled in the wire and shot it, then cut the wire around it. Without saying a word, he picked up the sheep and took it back to his Bradley. The man began screaming and becoming more theatrical, until I put my SAW in his face and screamed for him to shut up. We always said Iraqis don't speak English or Arabic, but they do speak gun. He shut up and left after we removed the wire, and eventually the military gave him money for all his lost and dead sheep. That night 2nd PLT came to the BIP, and we roasted that thing over an open fire, and I had hot meat for the first time since our arrival at the BIP. We slept well and woke up the next morning ready for a new mission.

Then we had to investigate a girl's college. Again, we were getting reports that the abandoned building was home to Al Qaeda fighters and needed to be secured. The imaging made the thing look more like a big house, but when we arrived, we realized it was a full campus. Unwilling to wait for backup and other resources, our platoon just entered and cleared the building. We found no evidence of IED materials, or any activity other than squatters. With the economy at a standstill for these people, many were living like hobos or drifters, and there wasn't a shelter that was off limits. People would sleep in shacks on private property and hovels alongside Trash Alley. It didn't matter, the apocalypse was here for them, their society had died the night we invaded and nothing was getting better.

What I found, however, was worth its weight in gold. I found a toilet seat. This small item was of such importance in this bleak world. We had not sat on a toilet seat to take a shit in what felt like forever, possibly months at that point. I took the toilet seat and attached it to my assault pack. Once I got back, I cut out the canvas of a field chair and attached it to the toilet seat with paracord. Since I had found the toilet seat, I demanded to use it first, and I was more excited than a kid at Christmas to sit and take a shit. Meanwhile the engineers had set up a barrier of Hescos around our new latrine area, and it seemed like the enemy took this period to decide to start mortaring us. I had my machine gun and body armor with me, but I had taken it all off to go use the bathroom. This was made worse because I had been eating MREs for months and was constipated. The mortars started dropping, and I had half a log out of my ass. I kept pushing like I was trying to break a speed record for giving birth. Out of nowhere I could only hear a ringing sound in my ears, and I was covered in dirt. I was face down in the dirt, my pants were around my ankles, and everything was blurry. I scrambled to find my gear and my gun. After I found them, I pulled up my pants and turned around. There was a giant hole in the ground covered in char, and the seat was gone. I began to cry and grabbed my shit. I walked out to see the medic running at full speed, and everyone was asking if I was okay. I finally just screamed out I DO NOT WANT TO FUCKING PLAY ANYMORE! The next question was from Davis: Where was the seat? I grabbed him on the shoulder and looked down.

"It's gone, bro, it's fucking gone."

A huge groan went out among everyone, and some guys started guttural screaming. This was fucking it for some of us, and we had FISTER figure out where the mortaring was coming from. We found the rounds originated from a nearby village and geared up to move out. People were shouting battle cries and hollering like Viking raiders. Our platoon raided the shit out of that fucking town more violently than any other town or any other battle I've been in. We were muzzle thumping every person we could, turning over homes to find extra weapons and contraband and beating people when we found it. We might have snapped, but we found the mortar, and the people that lived in the house next to it were roughed up and sent to Baghdad. They were probably some of the people the public saw tortured on the news. Human beings were tortured, beaten, and had their entire lives ruined over a toilet seat. When you live like a savage constantly in survival mode, the smallest things are so important. And when you lose those things, people become desperate and react in ways they normally would not. People died over the loss of that toilet seat. They should put that in the commercials and propaganda when they tell you what our military is doing to keep you safe at night.

A YELLOW BIRD WITH A YELLOW BILL
WAS SITTING ON MY WINDOWSILL
I LURED HIM IN WITH A PIECE OF BREAD
AND THEN I SMASHED HIS LITTLE HEAD!

CHAPTER 12

The Wall

The marvelous thing about a combat deployment is the loss of time. You start out keeping track of days and dates, but if you end up living outside the wire like we did, you lose track of all of it. You wake up and don't know if it's Saturday, or the Fourth of July, or your cousin twice removed's birthday. Rent isn't due, you don't have to worry about the percent your 401(k) didn't make this month, the baby doesn't need changing or the kid didn't bring home a shitty report card. You also don't know what the hottest movie that dropped is, or what song is at the top of the charts, or some media beef between celebrities or politicians. You do nothing but survive. You're so wrapped up in what's going on that you let go of society and become what you are in the moment. You are your best person, and you are the freest anyone has ever been. I've always said I was freer in Iraq, under constant military supervision and in a warzone, than any other place or time. You are outside time, and the people you leave behind seem to not miss the fact that you suddenly don't exist either. I missed the rollout of the iPhone, the creation of Facebook and YouTube, and all sorts of political and artistic achievements during that year and a half. When I returned, I didn't recognize music or language or know about current events. I was an outsider to my own country and hometown, but at the same time it was freeing and wonderful.

The one thing you never show in the movies is a dog dying. Never kill the dog on screen. You can show before, and after, just never the act. Even if it is natural causes, you don't break this sacred vow of cinema. Show the boy with the rifle, going to take care of his responsibility, because it is his dog. Show the after shot of the dog slumped over from behind. Hell, if there's a chance the dog makes it, then you take that chance, but never show the dog dying on screen. That's why the thing I had the most problem with was the health and welfare patrols, which we did on a weekly basis. Occasionally we would have to go around neighborhoods and shoot stray dogs. They were a menace to the local populace, and most had rabies or mange. They were breeding out of control, and Arabs do not like dogs. They consider them pests. Every time we went on a dog hunt it crushed my soul. We had to shoot adults and puppies. You did what you had to do, and some even could not manage it and it made them a bit crazy. The news posted a video of Marines throwing puppies off cliffs and shooting them with shotguns like skeet, but at the end of the day I understood why they did this. People go insane to get through things like that. That's why you never show the dog dying – never.

When we got back we would either immediately begin our details, go to our OP shift, or relax for a few hours. It wasn't often that we got down time, as we were running a full facility at the BIP and there was always something to do. SPC Larry from 2nd Platoon had been injured by having his eye pop out from a concussion blast. Since we were mostly patrolling in vehicles, Al Qaeda members were changing their tactics and using small grenades on parachutes to drop down on gunners in their hatch, and Larry didn't fully close his hatch in time. We were down so many people now that it felt like a constant grind between sleep, eat, shift work, patrols, and details. Yet we found time to do things like play cards or read a book. Trussel and I would play spades together, and we were the team to beat. I enjoyed the game, because it's more about strategy than about luck. If you have a good partner, you can guess what each other has well enough and place bets to screw the other team. This brought out the worst in SGT Zufall, who was a horrible gambler. He would gamble on just about anything with just about anyone and lose most of the time. Many of the lower enlisted, including myself, would regularly play spades with him and take him to the cleaners. But he couldn't leave those cards alone. He owed me close to $600 from our games, and he continued to play until our chain of command found out. No one could play cards with him anymore, and he was ordered to pay all his debts. I told him at first to forget about the debt, but apparently the snitch had given a record to our 1SG and senior NCOs so he could not get out of paying us. I almost felt badly for him, but then I remembered that he hadn't stuck up for me in my Article 15 hearing.

With the IEDs increasing on the road, and vehicles getting damaged daily, we decided it was better to start doing walking patrols. We wanted to keep our presence in the city and begin moving lighter and faster. By this point we had multiple HVTs slip through our fingers because you could hear our vehicles miles away. We would get some pop shots occasionally, and by now I knew how to differentiate if we were being shot at or not. Most of the time we weren't taking fire. The civil war between the Shia and Sunni factions had kicked off, and we would literally see people murdering each other in the streets. They would shoot at each other from their houses, then stop to eat lunch, then go back at it until sundown. If we were ever in the crossfire they would stop for the most part, as they didn't want us involved. We also had a few incidents of the Iraqi Army and Police fighting each other. They would drive out in the street and just start a firefight. Each faction thought if they killed the other they would be the group to control the new country, but this was far from the truth. The United States would never hand over power or authority to either of these groups. We watched our step around both and were convinced that those forces had Al Qaeda spies in them. We even heard some members of our company had bounties on our heads already, and if anything it was a badge of honor.

These foot patrols are how we stumbled on a local man in Tahrir who set himself up as a warlord. When Iraq collapsed, people lost everything. Their society shut down overnight and didn't pick itself up until after we left. People had no access to clean water, power, or government resources. In this hectic chaos many left the country, and others banded together. This man, instead of running or hiding, decided to go house to house and claim the entire street as his property. He murdered everyone on the street and gave away those houses to people who pledged to follow him. He then took over the entire neighborhood, at least three or four blocks, then moved into the biggest house on the street. He was running weapons and drugs and killing any IP that wouldn't work with him. He would capture their weapons, torture victims, and function as judge/jury/executioner.

We couldn't let this stand, and our platoon was ready to raid the house and kill everyone on the block. But we were told to stand down. We were here for Al Qaeda, and this man was apparently a good intelligence agent. This asset to the United States kept underage boys and girls as sex slaves and brutalized the people in that neighborhood. However, if he promised to stay just in the neighborhood and feed us intelligence, he was allowed to keep doing his thing. The veneer of society is thin. It's thinner than you think. Most people never notice, because there's too many things in the way of the veil. Terrible things happen, weather disasters, school shootings, protest and riots, but none of it matters. You can go to most towns in the United States and know there will be running water, electricity, internet, and a road to get in and out of town. We will always have a government to pick up the pieces and make sure you can still get your Starbucks while driving down the freeway to work. No one fears a total societal collapse in the United States, and our criticisms are for things we want, not things we desperately need to survive. What if there was no guarantee of the return of society? What if you thought your local government would never return, or the public utilities would not turn back on, or the banks would never reopen? The veil would lift, and your friendly neighbor Jim would turn into a cannibalistic sexual predator, or maybe the town would finally let its racist colors show and start lynching black people again. You would see your neighbors and families in their most base animal forms. There would be no question that civilized is only a state of mind, and it's very fragile.

OUT IN THE BATTLEFIELD, COVERED IN BLOOD
THERE LIES A RANGER, DYING IN THE MUD...

AIRBORNE...

RANGER...
DIE DIE

SHOOT, SHOOT, SHOOT—SHOOT TO KILL!

WHY WON'T YOU DIE?

CHAPTER 13

MERRY MEN ARE WE

Once again, another lieutenant was passed down from battalion to our company, and he went straight to Troublemaker Platoon. He was still in charge of the Blood Angel contingent and Trouble, so he would split his time between the two groups. Still, we were mostly all at the BIP by this point, and he didn't need to go far for either part of the platoon. This was the first officer that was intimately involved in lower enlisted affairs. This guy would walk with us, eat with us, and joke around. He was trying to really become one of the guys. He ended up being our longest-tenured lieutenant, and I would argue the best. We would talk shit to each other in the tanks, and he could take criticism just as well as dish it out. The largest factor that helped him was that he listened. He didn't act like some West Point jagoff who knew better than you did, even though you had been in country longer. He knew the line, and watched, and obeyed our PSG and squad leaders. He kept his side of the job as going up to speak to the brass and the local leaders. He kept himself as the face of the platoon but, before giving a final answer, he spoke to his leaders. We ended up hazing him in the group to show our affection for the man. We did what the tankers called the "Dancing Tank": We stripped him butt naked, taped him to the barrel of a tank, and spun the tank barrel around while spinning the bottom in the opposite direction. When he had thoroughly puked his guts out, we then taped him to a chair and sat him in front of the XO's room, knocked on the door, and ran away. The XO opened the door and looked at LT Sparks and said only: "Welcome to BCO, Joe." We laughed and took him back to his room and, from that point until he left, we knew we had the right guy for a bunch of Troublemakers. He wouldn't snitch and could play just like the boys.

Now it was time to get to work, and the work was dirty. Our platoon seemed to be the best to task with the issue of an IP general who was found to be working for Al Qaeda. We found out because one of our first interpreters had been kidnapped by terrorists, then executed after feeding them information for a week. This terp was found out by some investigations unit and ran away to the terrorist cell, who decided he wasn't worth the trouble and executed him. We found his body, burned and shot to pieces in the street in downtown Baqubah. This led to a huge investigation that resulted in the first IP general to be tried and executed. They also found evidence that he would feed us false intelligence and send us on wild goose chases to allow terrorists time to bury IEDs. We would go to the locations the general suggested, only to come up with no intelligence and then run into an IED on the way back to the COP. I ended up being placed on the security detail with LT Sparks for these events. The sight was intense, and shocking. I had only seen gallows in old Western films and read about them in books, but now I was staring at a platform surrounded by hundreds of people screaming at the individual on the end of a rope. I was there to keep the peace, but at no point was I going to fire my machine gun. The crowd was too big, and the idea that I would be able to return fire if someone in or near the crowd decided to do something was insane. Security was hanging by a thread, but apparently the people in the area were angry enough to allow this to happen.

I watched as the man was placed in a hood, the rope was put over his head, and he was dropped from the gallows. His body swung below, and the people cheered. Some tried to rush the stage but were pushed back by other security forces. I watched as the man twitched for a minute. I ended up witnessing this event another two times, as the other IP generals that replaced this man did the same thing. I started becoming numb and pulled less security each time. I figured with the chaos at the event, if someone was going to do something, I would need to be more reactionary than prepared. I would need to be prepared to take a bullet for my LT, rather than try and fight anything at the hanging. Because of these events, we began to distrust all institutions and those that led them.

The next thing we learned was that people were being used to place IEDs because they were poor and destitute. The lack of commerce caused people to take jobs for anything, from anyone who paid anything. Situations were occurring such as a man being given a bomb to place in a random hole while walking in the morning. You had an unlimited number of people willing to do anything, and the United States was doing nothing to help these people. The level of intelligence among the masses was so low that you might even believe them if they told you they didn't even know the thing they were paid to bury was a bomb. We had to do something to get people on our side, so we began adding missions to our presence patrols. We would give people food, water, toys, and random goods. These events were hectic, as we had kids try to hand us grenades and walk away. In one instance a child was blown up by the grenade he was carrying because he didn't make it to the front of the crowd in time. Nothing we did made a difference, and we were still perceived the same as the Al Qaeda fighters. I didn't understand it then, but I certainly can now. I can see how an occupying force in your neighborhood that is constantly attracting bad attention would be blamed for these events. It didn't help that we would rough people up while searching homes or imprison people for the slightest infractions. They would bring their dead children to our COP and insist we were at fault, because Al Qaeda had come only after we arrived. We shut their lives down and didn't do anything to create a better life. You can't blame them at all.

Six high-rise buildings stood out against an open field of grass. We called these buildings The Projects. They were at least six or seven stories high and looked identical. There were wires hanging back and forth between windows, with laundry draped over the lines. The middle quads of the buildings were flooded with human waste and water. We made a first attempt to contact the people in the buildings to see if they needed some help. We drove up with our Bradleys and HUMVEEs and called out to the buildings using the terp. It was then that a few men walked out to tell the terp that it wasn't good to be seen talking to us, and that we needed to go. We then tried to give them offerings of some food and toys, and they refused, saying they could be killed if they were seen accepting these things. We left them around back so the residents could make use of them, but we also knew that was not enough. As we left, we watched the people descend from their rooms like rats onto the food and gifts, fighting with makeshift weapons and pushing each other aside. The next time we drove by we saw a dead family left in the open where we had left the gifts. We called it in, and they told us it wasn't our place to take care of this. It wasn't until a few weeks later that we returned. Battalion finally saw fit to send us with some resources to encourage these people to join our side. We drove out with our LMTV stocked to the top with rice and other goods. We were meant to make this some sort of peacekeeping and health issue. Doc McCollough was going to try to see the locals and give medical aid, and we would pass out food.

Yet when we drove up to The Projects again, we saw the same men walk out. While they were speaking to LT Sparks, we moved into the quad areas and began setting down the food, medical supplies, and toys. I looked up and noticed everyone was out on their balconies, staring down at us. It was ominous, and no one spoke a word for what felt like a lifetime. I knew this sensation. It felt just like back on Camp Red Cloud in Korea. I slowly looked at Webb and said, "Don't shoot anyone, for any reason." He looked at me confused, and I repeated myself. "If we start shooting, this place is going to turn into utter chaos. We must leave here as soon as possible." Before he could say anything, people were already surrounding us and began grabbing us. Davis started yelling to get off him, and some other guy was screaming. I yelled out to run just as I heard Webb's SAW let out a burst. "FUCKING WEBB what the FUCK?" We began running, and even though the quad wasn't that big, it felt like it took forever to move through the crowd. They were grabbing and pushing us like zombies. We ran until we finally were out of the building, and I saw Davis move to the side. Webb and I began shooting at the crowd. The first few people fell dead on the ground, and we looked stunned. I had hoped this would stop them, but we saw the rest of the squad running from the other side of the building and then the crowd rushed forward. We let out another burst, and then the Bradley opened with the coax and cannon. We heard SSG Shaefer on the mounted squawk box, screaming for us to mount up. I dove into the Bradley with Webb and Davis jumped in the HUMVEE. We could still hear the shooting as we drove away, and until we left we never went back to The Projects again. Anytime we were around them, we made sure to stay far enough away that you couldn't make out who was inside.

A few days later the mortars were kicking back in full gear, and a firefight broke out at the gate. I rushed to the front, where I could see multiple men moving across the road in full black. They had guns and were shooting. I returned fire and dropped a whole nut sack of rounds and reloaded. At this time 1SG Colella noticed me, saw that I could get pinned down, and came out and returned fire over my head. I was astounded that someone of his rank would put himself in danger for a shitbag like myself. He screamed at me to reload, and I hurried and began shooting and laughing. Then from out of nowhere I looked across the street and saw SGM Harris, crouched behind a wall while his detail fired above his head. He began screaming at me, but I couldn't make out what he was saying over the sound of gunfire. Even 1SG was asking him to repeat himself. I then looked at his lips and understood what he was screaming: that my boot laces were untied. I looked down, and sure enough. I looked back at him and screamed, "We are fucking getting shot at!" He looked upset, but I didn't care. Who the fuck was this idiot to freak out over my boots while we were taking fire? 1SG Colella looked over confused and told me to continue firing. I kept going, and we killed the guys at the end of the street. 1SG told some other guys to go conduct BDA, and the SGM began crossing the street with this angry look your mother has when you don't listen to her. He started screaming that he had ordered me to tie my boots but I hadn't listened. 1SG looked quickly at me and told me to tie my boots, then told SGM to follow him. He said he would "take care of it" and they had more important matters. I moved down the street to collect BDA and talk shit about what had just happened. I was thinking about how neat it was that I was just walking by myself down the street, something I hadn't been able to do for months. As I got closer to the corner, I could see people running back in a full sprint. I called out for what was going on, and they screamed, "Spencer, he's been shot! Spencer's fucked up, go back!" The next word was chilling: "SNIPER!"

We jumped into the ditch. My clothes and gear were now covered in shit. I could smell the sewage covering my body. We high crawled back to the front gate and bounded inside. Spencer was being moved to the rooftop, and people were screaming. I got one look at Spencer before turning to pull security. His hand was shaking, he was in tears and breathing erratically. I knew he wasn't going to make it, but that rooftop had to be secure for the bird. I traded my SAW for a 240B (bigger machine gun) and posted up where I had the most cover over the greatest number of houses. I was flicking them off while aiming down my sights. By the time the chopper landed and pulled off with 1SG and SGM inside, I was sure Spencer was dead. I looked down, and my laces were still untied. I went back to my rack and changed all my clothes and smoked a cigarette on my cot. Fuck that POG.

I WENT TO WAKE THE CAPTAIN
HE WAS LYING IN HIS BED
I WENT TO TURN HIM OVER
I FOUND OUT THAT HE WAS DEAD...

CHAPTER 14

Dazed and Confused

The call to prayer rang out throughout the city as the sun rose over the huts and buildings made of mud. It looked like Agrabah from *Aladdin*, except you could smell the shit and trash. You could feel fear and death in the air. We were ready to descend on Baqubah in a way that hadn't occurred since before we arrived. I looked out over the city and could see the bell tower as well as the hovels that spread across the terrain. I lit my cigarette and checked my gear. It was going to get nasty today, as we were going to clear this city, block by block and house by house. The entire company was coming down on Baqubah, and we had even used some POGs to pull security on the COP while this operation went down. I could see a convoy of HUMVEEs and Bradleys moving in front of me, beginning to peel off down the side streets. The first few houses were uneventful, and we found nothing, but as we made our way through the city, reports started popping up from other platoons that they were finding extra weapons and IED materials. The next few houses we found extra weapons as well. We were processing people and sending them back to the rear to be moved to other locations when a shot rang out throughout the city. I could hear the rotors of Black Hawks and the grinding of tank tracks moving. If you haven't heard tank tracks moving, it is very specific. You can tell the difference between a small track vehicle and a tank. They sound like they're breaking and snapping metal, and as the vehicle accelerates it sounds as though a monster from hell is groaning. I thought the first people to encounter tanks on the battlefield must have been terrified, as I know I would be if one of these vehicles were tracking me. The shot rang out again, and I could tell it was the sniper. The radio began blowing up with yelling, and I could tell they were coordinating vehicles by the call signs I could hear. SSG Haas told me to get down and close my hatch. I did this just in time to see a man drop a grenade with a parachute from the top of a building next to us. I closed the hatch and pulled the lever to lock the door, then vibrations rocked the vehicle. SSG Haas was screaming over the radio, and SGT Crowley grabbed me to check if I was okay. I asked SSG Haas if he wanted me back on the gun and he said no. He ordered the dismounts to charge the building, and SGT Crowley led the team inside. I could hear shooting and knew they had the guy pinned down. After another grenade went off it was quiet. SSG Haas waited for some chatter, then began screaming for a report. You could hear SGT Crowley come back, and SSG Haas confirmed they got the guy. I was ordered to reopen my hatch and we drove forward.

Then the bell tower began ringing and the call to prayer seemed to get louder. 1SG Colella was driving up to us, since we had a terp, asking what they were saying. Snake told us they were calling all fighters to defend the city. Then over the radio you could hear SFC Malo requesting to use the main gun. We understood the ROE (rules of engagement) at the time were to leave religious buildings alone, and the tower was attached to the mosque. A few teams now had identified that a man was in the tower, actively shooting at us. 1SG Colella told the tanks to hold and called up HHC for the sniper team. We would have to surgically take him out without destroying the mosque. Our sniper team moved through the city under cover of our HUMVEE until they were able to set up on a tall building with a line of sight to the tower. They moved inside the building with another squad for security and identified the target. The deafening silence was finally broken by a Barret .50 cal. We heard the radio call out and I could hear it was a hit. They got the guy. We continued clearing the city until we made our way to the mosque.

We couldn't go inside, we had to respect this building, but some deal was worked out between 1SG and the imam to bring the body down to us. When it came outside I could see why: This man looked East Asian. 1SG took the body in his Bradley and moved out to FOB Anaconda. It was confirmed: The Korean snipers were a thing, and that meant there was another brother out there hunting us.

When we got back to the COP, I saw someone I had not seen in a while. SPC Milligan was the NBC soldier who controlled our Pro Masks and other chemical items issued to the company. He had tested and issued our equipment before we left, and I had not seen him since. I assumed he was tasked out at battalion or some other area away from the fighting, but he had been somewhere else these last few months. For reasons that will become obvious, he was sent to training while we were deployed. Milligan was a nice guy, but stupid. He couldn't breathe without his mouth hanging wide open, and most of what he did say sounded like a fifth grader's best attempt to pay attention to the question in the first place. He was weird and didn't hang out and drink either. Because of this I think they sent him to UAV or drone training. He was good at technical things, and the lack of sociability or common sense could be deadly down range. Yet he completed his training and knew that little unmanned flying robot inside and out.

He had four big boxes in the LMTV behind him and each had a drone inside. He brought them out for everyone to see and gave us a small rundown on their capabilities. The one thing he did make sure to tell us was that no one, not even 1SG, was allowed to use these UAVs without him physically being there or on the radio to guide the user prior to the fact. He really showed his colors that week when we started taking mortar rounds and he flew the UAV to identify enemy targets and allow the Bradleys to flank and kill them. Yet soldiers get bored, and once Milligan left for the FOB, we immediately pulled them out and began playing with them. We were putting them together when the XO came out to the roof and said he had missed Milligan's demonstration and wanted to see them fly.

He grabbed the controller and was setting it up, when we reminded him that he needed to get Milligan on the radio. He told us that technically being the company officer it was his property, and he had been to college so he should be able to figure this out. We laughed and let him play with the controller. Eventually he was set and told us to throw the UAV. It flew straight up and crashed on the roof before takeoff. The birds were meant to fall apart when they landed, so we freaked out, thinking at first that we had broken it. The XO told us that's what happens and to put them back together. We tried again, and this time the bird flew. It started out low, then lifted higher. He was using the controller to see what the ground showed and put on the different filters for us. After about an hour of flying, he decided it needed to RTB and tried to bring it back. That's when his face went pale.

"It's not coming back."

"Did you try using the joystick, sir?"

"Yes, I fucking used it!"

"Don't get mad at me, sir, I'm just trying to help. Should we call Milligan?"

"Don't you dare call that little fat fuck!"

"Okay, sir…"

He began beating the controller until SGT Crowley came up. He reminded the XO that he shouldn't have touched the UAV without Milligan. At this point we were all giggling like schoolgirls at the XO, and he was yelling at us to shut up. Eventually, after trying everything, he relented and got Milligan on the radio. He wanted to play it off that he was setting it up and hadn't been flying the bird yet. Milligan ran through the prompt, until he asked if the XO had given the UAV RTB coordinates. He started to stutter, and Milligan repeated that it was important to give the bird RTB coordinates in case it locked up. This would have the thing automatically start coming back if the battery hit a certain percent. The XO was stuttering more, couldn't get a full sentence out, and we heard Milligan come back over the radio. "You already threw the bird and were messing with it, huh, sir?" The XO relented, and immediately 1SG popped up. "Looks like you just bought a UAV, SIR!!!!" We began laughing, and he was red with embarrassment and anger. He told us we were going to mount up and get this thing. He started running downstairs screaming that he needed a detail. 1SG Colella stopped him in his tracks: "I'm not sacrificing good men for your fuckup." We stood down, and the last we heard anything about the UAV it was straight on a course for Somalia.

Then I was standing at the front of the gym, back in formation for a twenty-one-gun salute. I heard it all again: The fucking people laughing, and the people who didn't even know him talking shit. Then the organs started playing, and the priest was going on and on about how this young soul was taken too soon and would be in heaven, even though he forsook his religion and went full atheist. Then the fucking wailing from goddamn people who didn't even know him. I lifted my rifle, this time with tears rolling down my cheeks. I fired, and brought it back, three times. I was then told I could go inside and make my peace, and there hung a picture of Seiger. I saw the podium with the rifle, helmet, boots, and dog tags. I couldn't believe this had happened.

Days earlier, Seiger had been driving around the COP in his Bradley during a rainstorm, and the vehicle became unstable on the side of a ditch and flipped over. Seiger was trapped in the driver's hull because the Bradley was upside down, and the dismounts had their gear in what is called the Hell Hole, the small tunnel that connects the driver and the back of the Bradley. The rucksacks were not secured and slammed down the Hell Hole, trapping him. We got the call at the COP and immediately mounted up. It was down the street, and we were there in less than a minute, and people were standing around looking. Barton couldn't take it anymore and jumped in the water and swam down to the back of the Bradley. He opened the back door and brought gear and rucksacks to the surface. I heard SGT Zufall and Fuller yell at him to get out of the water. He told them to fuck off and dove back down. I stayed in front of him on the bank pulling security, and SSG Haas ran over to help him grab items from the water. He couldn't reach Seiger, and he lay on the bank of the Tigris out of breath.

It seemed like hours before the wrecking crew and tanks showed up. They pulled the Bradley from the river, and I watched as the lights illuminated the vehicles. People were shouting and running around, but I stayed focused on security. The last thing we needed was to be caught off guard while recovering Seiger. As the crane moved the vehicle to dry land the back door released a heap of equipment, and then Seiger's lifeless body fell limp from the back end. He hit the ground, and Barton ran forward to begin mouth-to-mouth. It was too late, though; he was dead. We found out that our supply sergeant had a full connex of emergency canned air for all Bradley crews just for this scenario, but he had refused to sign them out. He had some off-putting excuse about how we would most likely abuse them. He was immediately reassigned, and with good reason. If the company came back to the FOB he might have been jumped, and we'd have another Malitz incident on our hands. Years later I met him at another duty station, and he hadn't changed. He was going on about being able to game the system and use his position to his advantage. I couldn't bring myself to beat his ass, because I felt if I let him get to me, I would be doing Seiger a disservice. Who knows if ten minutes of emergency air would have helped, but we will never know. After all this, at the end of the deployment the NCOs who were part of the recovery mission were given Bronze Stars with valor, and Barton, the man who braved the Tigris River to get our friend, got some stupid ARCOM (Army Accommodation Medal).

As I lowered my M4 for the 21-gun salute and went inside the gym, the picture in my head was made worse by reality. People were fawning over Seiger's picture. People who didn't even know him were bawling like he was their closest family. I heard one female soldier cry out that he was looking down on us in heaven now. I thought, "You stupid bitch, he's giving you the bird while waiting for us in hell." I walked past the chaplain who was praying for people in full tears. They didn't know him. I even saw contractors crying and thought this was some theatrical bullshit to fucking get out of a detail or work. I said my goodbye and left; I was raging inside. I walked over to the PX on FOB Warhorse, to buy another few cartons of cigarettes, and they were completely out. The POGs on the FOB would buy them all within a day or two after the PX was restocked, and they didn't have one fucking carton left. I walked back clenching my fist and got in my gear to go back to the COP. I just wanted to go back home, and home was the front line. I was sick of being around polite society. The chaplain walked out and saw me smoking a cigarette and sneered. He told me I needed to be fifty feet away from the building to smoke. That's when I snapped. I began cursing him out and flicked my cigarette at him. SSG Haas ran around the corner and pushed me down and turned to the chaplain to make up some excuse about how I was grieving. I just stared at him in anger and hate. We didn't have time for chow or showers, and we drove back to the COP.

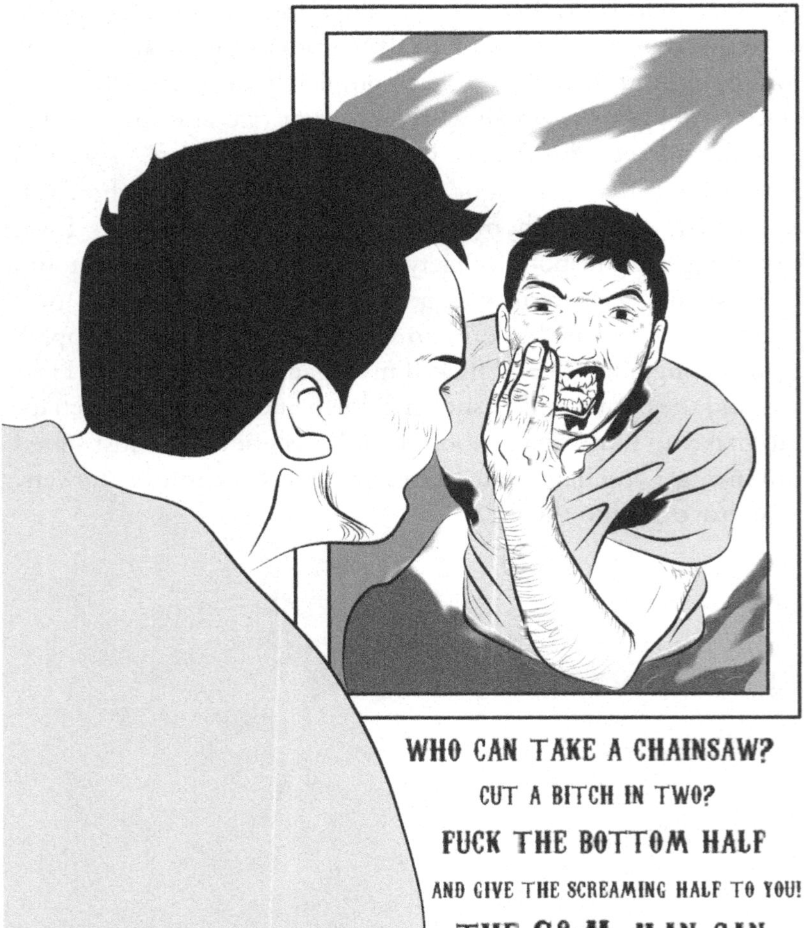
WHO CAN TAKE A CHAINSAW?
CUT A BITCH IN TWO?
FUCK THE BOTTOM HALF
AND GIVE THE SCREAMING HALF TO YOU!
THE S&M MAN CAN...

CHAPTER 15

Now It's Time to Say Goodbye to All Our Company

We spent at least four months at COP Bone Disney before we went back to FOB Warhorse. With the number of people we had lost so far, we were exhausted and lost. We couldn't get proper gear or people, and it was beginning to show. A break in the circle of survival finally came when we were allowed to go back to the FOB for one weekend. There were some rumors that we might be getting some replacements, and our time at the embassy was finished as a company. Because of this we could now cycle platoons out of the COP and have more people for details and patrol rotations. We drove back near the end of the day, and everyone was salivating over the fact that we would get to eat hot chow. It had been four months since I had had a hot meal, four months since I had felt water touch my skin from a shower, four months since I had worn a clean uniform. I stank so bad I didn't even notice anymore, and only God knew what I looked like at that point. We heard it from PNN that the real reason we were going back was that no one had gotten a haircut in four months and SGM Harris was pissed. I don't care if it was true, but I believed it. That fuck would care about something so stupid.

As the sun started to set, we pulled up to the FOB and noticed all the lights were on. You could see the damn place from fifty miles away. It didn't matter, we wanted food. We pulled up to the chow hall and started lining up. We saw soldiers begin walking back from the front of the line. 1SG grabbed one of the privates and asked why, and he told him the chow guard said we were too dirty. The animalistic rage began to burn in our minds. For the last four months we had suffered and lived like animals. 1SG snapped; he locked and loaded his M4 and began walking to the front of the line. Our NCOs saw this and began to chase after him. The chow guard saw this, ran inside, and came out with an SFC. As soon as he saw our 1SG he locked up, and you could see he was scared. 1SG assured that man that nothing was going to stop us from eating. We had gone through the shit while these POG fucks lived a life of luxury on this FOB. The only thing that was going to stop us from eating was a bullet. The chow NCO quickly relented, and we all began to flow into the chow hall.

For the first time in four months I felt the shock of AC. I began to shiver uncontrollably and looked around to see the greatest sight in my life: giant blue crab legs, steak, shrimp. This was like Charlie walking into Willy Wonka's factory for the first time. I was singing the Gene Wilder song in my head when I was snapped back to reality by an Air Force guy complaining: "Surf and turf again!" He sighed like this was the most disappointing thing he'd ever experienced, like someone had just kicked his puppy on Christmas Day. I found my hand on my knife, shaking, and Joe grabbed my shoulder. "Bro, it's not worth it." The guy turned to see what had just occurred, his eyes got big, and he immediately moved out. We sat down and ate like kings.

However, this was quickly dashed as well, when mortars began landing near the chow hall. People began running and screaming like in a 1950s horror film. We sat there, unaffected. We continued to eat, and I grabbed the Air Force guy's tray for more food. Not one mortar hurt anyone, and when we were done, we left back to the Chus.

That night I had my first shower in four months. I didn't even recognize the creature looking back at me in the mirror. I didn't see the human being beneath the dirt and grime. That was the best shower I've ever had in my life. Eventually the weekend was over, and we had stocked up on goodies and cleaned off. We got haircuts and tried to have our clothes washed, but they were so soiled the laundry wouldn't take them. They just issued us all new uniforms and undergarments. We went back to the COP well fed and rejuvenated, but I think they just wanted us to look presentable for what was coming down the pike. If anyone had seen the state we were living in, the command would have looked bad, so we needed to be dolled up for the dog and pony show.

Back on the COP, on the wall in the common room was a large blackboard. Maps, GRGs, HVTs, and other information were taped to it and the wall around it. There would be updates each day on mission OPORDs: details, listings, times, and other pertinent information. I walked in that day to conduct RTO detail and log high priority events, and noticed the board was blank. What I had previously enjoyed about this board was that, along with the intel, I had drawn a map of the place with chalk on the board. We nicknamed the COP Bone Disneyland, and each OP was named after a Disney ride. Splash Mountain was overflowing with sewage water from a busted pipe. The commo room was the Epcot Center. Thunder Mountain was the OP that had been hammered by mortars too many times and was falling apart. Space Mountain was a dark hallway (which we dug) between the buildings to connect sleeping quarters and the main rooms. The finale was the "recreation area" where our porta shitters, card table, and burn pit for trash were; it was called It's a Small World. We weren't the guys who did everything by the book or used the same boring ass call signs and mission names. We were fucked-up professionals who hated our job and everyone else.

It surprised me that everything had been erased, and before I could ask why, 1SG came in and told me he had new information for the company. I was to call all call signs on the net to RTB for the COP. I did as he asked, and while we waited, he told me the news. We had finally been given a company commander. He was already changing the landscape of the company and coming in swinging his big dick around. No longer would we be Bone Disney. It was now COP Adam, and these childish names would be back to OP 1,2,3. Even our call signs had to change. We were now Red, White, and Blue Platoon. Eventually everyone would come around to CPT Austin, but this would be the one thing I held against the man. He didn't initially understand who he was dealing with. For Christ's sake, for at least four months, instead of the U.S. flag I had worn a pirate flag on my shoulder, with a Jolly Roger and Troublemakers on it. Not until the SGM saw it and took it away did I stop wearing it. The running joke about rank was that you would never get to SGM unless you had at least one divorce, one DUI, and one field grade Article 15.

So as per usual when a new CO arrived, things began to change. Maybe there was a thought to add some sort of validity or respectability to our unit, but that was never going to happen. The average soldier who was new to combat deployments was jaded, never mind the guys who had multiple deployments. We had UCMJ stacked against almost an entire platoon, half of the company was on STOPLOSS status and were supposed to have gotten out of the Army before we deployed, and now we were stacking up injuries and deaths left and right. We didn't want legitimacy; we didn't want to be the premier unit of the Army. Most guys just wanted to go home, unlike myself and Joe who just wanted to kill as many people as we could before we left. The upside was that our shifts were restructured. We would only do six hours on an OP and were limited to hour-long patrols during the day due to the hundred-degree heat. The week prior I had stayed awake for seventy-two hours straight because our platoon went on a mission and there was no one to watch the COP except our squad. When the company came back the SGM came up to my tower and caught me sleeping, while standing up, with my eyes open. Instead of punishing me, he demanded I be replaced. I was walking back down to the sleeping quarters when SSG Embry screamed at me to go back to my post, and the Brigade SGM, told him to fuck off and that he was going to man my post until someone else was sent up. Our new CO was able to go to battalion on our behalf and request platoon leaders. He also was able to request more property, which was good for us, because at this point, we had been riding around Iraq in HUMVEEs that had no up armor. Some still had the tarp doors that are on stripped versions for training. There were broken weapons like an MK-19 that was destroyed when SGT Crowley couldn't get it to function correctly and threw it out of the window during a firefight. The actual parts of a company that should work were working for the first time since we had left Fort Hood. Yet, at the same time, fuck him for taking our spirit away.

Of course, the installation of a new king shit officer requires the dog and pony show. The "Big Army" came out to our COP to see how we were doing. Generals Petraeus and Odierno were two big-shot generals from III Corps, and both would go on to be the commander of all operations in Iraq. There was a line of HUMVEEs that were completely armored, outfitted with the best equipment and anti-IED technologies and the most advanced weapons, like the CROW System. It looked like Kim Jung Il's convoy, and when they stepped outside at least a platoon of people surrounded them. The cavalcade of circus members made their way through our compound, and it was now louder than it had ever been prior.

There are certain ways to live when you live in a patrol base in combat situations. Noise and light discipline are key, and they were being trampled on for the sake of this parade. I noticed we weren't shot at or mortared once, and the whole city seemed to be on lockdown. Of all the places this eventual convoluted bullshit of fucks could end up, they ended up going to my OP. I began to talk nervously about our range cards that I had made and showed General Odierno the blast shield and mounted gun at a window. I remarked how we got shot at every day from a building close to that window. He looked under my blast window to see where I was pointing but was not wearing a helmet. I freaked out and grabbed him to pull him back. His entourage ganged up on me and began to yell and put me on the ground for touching him. He stopped them and grabbed my hand, pulling me back to my feet, looked me in the eyes and said, "You were right. I was not wearing my cover, and remember you can always check up." I opened my hand to find a coin in it. My CO ran up after the event and asked what I had done. I said I didn't know, but I got a coin. He laughed, but after that any time Odierno went on a patrol I was requested, and I patrolled right next to him as a team leader. These events are more important to me than to them. They probably don't even know the impact they had on me.

As if the fanfare of bullshit couldn't get any more ridiculous, CNN decided to come to our COP. They wanted to document what we were doing in Baqubah. As they filmed, our CO led the way and explained what the rooms were for and how we were living. It was like a fucked-up episode of *House Hunters* starring two fucked-up dads with a hundred and fifty shitty stepchildren. The cameraman was in lockstep behind our CO, following him like he was a movie star or some shit. As they made their way through the halls and asked CPT Austin questions, they moved near a room that we never used for anything because when we had cleared the building, we had thrown a grenade inside it to take out some terrorists. The walls were covered in soot, and it was completely black, which to everyone else meant nothing. To me, I now had four huge walls to create my next masterpiece. Between shifts I drew anything and everything I could. Some people came in while I was working and would add small scribbles or doodles, but most of the compositions consisted of animals with huge dicks. I drew the combat monkey I had on my turret shield, and of course a cyborg baboon with a huge dick. The masterpiece was opposite the door as soon as you walked in. I called it "The Ninja Turtles Gangbanging April O'Neil," because that's what it was. The CO and cameraman walked to the door, and CPT Austin opened the door wide enough to get a glimpse of what was on the wall. He quickly closed the door, but the camera operator saw and was trying to get a shot.

He missed out just in time for CPT Austin to proclaim the room highly classified.

After the prancing pony tour of our home, he walked back to the room and stood in awe of the pure smut and filth on the wall. He declared we were all fucked in the head and walked out. As he left, I asked if that meant we could keep the art. He just kept walking. I think it dawned on him at that point what kind of soldier he was dealing with in the Bonecrusher Company. The funny thing was that Katie Couric was the one who was supposed to interview us, and some bullshit event happened that kept her from arriving. So, if you're reading this, Katie, I made the mural just for you.

POPPA TOLD SALLY NOT TO GO DOWNTOWN
TOO MANY BONECRUSHERS HANGING AROUND
SALLY GOT THE ASS AND SHE WENT ANYWAY
SHE CAME BACK HOME THE VERY NEXT DAY
THREE MONTHS LATER ALL WAS WELL
FIVE MONTHS LATER SHE BEGAN TO SWELL
NINE MONTHS LATER OUT IT CAME
IT WAS A BABY BONECRUSHER SWINGING A CHAIN!

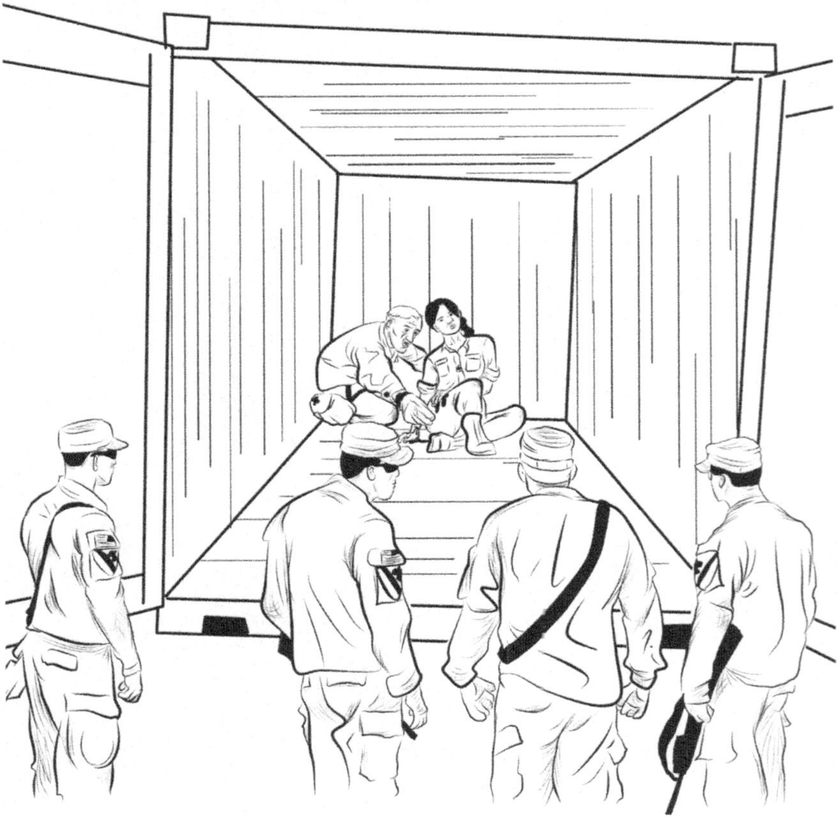

CHAPTER 16

Rivendell

Davis and I pulled up to the fuel pumps on the FOB. We were now in a rhythm of putting fuel in the vehicle every time we returned and, depending on the time of arrival, moving to the motor pool for repairs. Along the way we would stop by the burn pit and throw out whatever trash we had, making sure to dodge the stray bullets and ordnance ignited by the fires. This time we were fueling with the tankers, and we checked in with the fuelers to shoot the shit and listen to a few stories. We traded names with one guy and his face lit up. He said he knew us; we were infamous as "the Wild Boys of the Battalion." I thought he was pulling our chain, but another fueler confirmed it. I began laughing, but Davis was more cautious about the implications. He told me that if we were known that well that important people had us on their radar, he didn't want that. I told him he couldn't control it, and that I enjoyed the spotlight. We moved back to the motor pool to begin working on the HUMVEE. I then noticed we had new armor for vehicles. We were now supposed to replace the old doors and put up-armor on the vehicle. Each piece weighed about eighty pounds and was full of explosives to redirect IEDs if they hit the doors. We outfitted the vehicle, then tried to drive it back to the Chus. The vehicle wouldn't move. I brought a mechanic over, and he quickly identified that the chassis was too weak to take the new armor. We wouldn't be able to drive with the weight. We had to take everything off and go back to SSG Haas. He began cursing and told us to wait while he spoke to 1SG. We were then told to just put the armor on the doors, and that now we would have to get new vehicles.

We laughed it off as another complete failure of the Army to efficiently plan anything. The next day we were given a new vehicle and told we had to outfit this with the new armor, and a THOR. This equipment was supposed to stop radio/cell-activated IEDs and keep us safe. We put the armor on, but even then we could barely move. My turret was so heavy I couldn't get the crank to move, and if I didn't lock it, I would spin around out of control. It wouldn't be until after our next rotation that we finally received the proper equipment that could handle the new armor specifications. We would have to make a trip to Balad to get our new shit. What bugged me was, for one, that we were letting go of history. Whoever had just died in that seat when we first arrived was being let go. No one was going to remember him, and I didn't even know his name. The next problem was that the anti-IED equipment was installed by civilians, and we weren't allowed to touch it. We were going to be the test dummies to see if this shit worked, and if there were any problems, we were supposed to RTB immediately.

This caused every crew to have to RTB over the next month in the middle of missions, patrols, and changeouts. The damn thing would shut off for no reason, and we'd call it in and go back to the FOB. People were using it as an excuse to go back to the FOB and get shit from the PX and out of missions. We were already undermanned, and this wasn't the right time for people to be playing hooky. I also noticed that every POG unit on the FOB already had new armor, and THOR units. They had vehicles that could accommodate the weight and electric turrets that turned with a joystick. As per usual, the infantry had to suck it up and get the shit gear and wait our turn to have decent equipment. I noticed that the route clearance crew, the only POGs that left the wire, had a new vehicle, the MRAPP. I thought this was complete bullshit, as this vehicle had more room and armor on the bottom. However, we quickly found out that this was also a test vehicle, and people died depending on that lower armor that wasn't built to specification. I believe the company was sued by the Army later for this mistake, but it didn't stop them from buying and issuing them out. We went to Balad to get our new vehicles, and then a bombshell was dropped. 1SG told us we were below sixty percent strength as a unit, and battalion had been sending up wrong numbers to the Big Army, and that's why we weren't getting the desperately needed replacements. There was a platoon of privates in the rear just being held back when they could have been shipped forward. Because of that another unit was now taking over operations in our area, and we would fall under 2ID 5-20 BN. We were instructed to report to this new SGM, and they would ensure we got new vehicles and replacements. We weren't allowed to leave Balad until our replacements showed up.

It felt like forever waiting for these guys. These were the same Chus we stayed in while waiting to get orders for FOB Warhorse, and people were yelling at us regularly for not wearing a PT belt to chow, or not immediately jumping to a bunker when the incoming call was sent out. You'd never see a mortar or an attack, but they freaked out like it was D Day and we were going to lose the beach. Finally, after a lifetime of waiting, our replacements showed up. There was SPC Erk, SPC Puffer, SPC Ward, Cummings, Valverde, Miller, a new medic, Doc Nguyen, and a new interpreter we called Steve. This terp was extremely effeminate, and immediately he and I became the butt of jokes saying we were gay because we shared the same first name. I blew it off and understood that there are cultural differences, and Steve seemed to care about his job and the state of affairs. He wanted to fight for Iraq and wanted a better society. I spoke to him in depth about cultural issues that were taboo to honestly learn more about the people we were supposedly protecting. I picked up a few phrases and gestures.

Puffer and Valverde went to our squad, and the rest were dished out among the platoons. We had since lost SGT Fuller due to his misfire incident during mid-tour leave, and he wouldn't show back up until the last two months of the deployment. We were restructured again, and Stephens was added to our squad. It was now SSG Haas, SGT Crowley, SGT Zufall, SPC Puffer, SPC Case, Davis, Melsha, Stephens, Webb, Valverde, and me. SPC Case was a weird one. Even though he was in our squad, he had some shit going on back home, and SSG Haas did him the favor of letting him continue to drive tanks with the Blood Angels. However, this got messy because they were officially separate now, and he would still be on our schedule. Sometimes we would get back from the COP and he would go straight to the motor pool to work on the tanks. Other times, he would go straight on missions because they were riding out at the COP. I didn't particularly care for Case because he was dirty, and because of him 1SG was always getting on Davis and me about our living quarters. When he left on his mid-tour leave he left his computer in our Chu, but his power cord was nearly destroyed and almost burned down the Chu when it started sparking. SGT Zufall had similar arrangements, but he had the privilege of being an NCO and refused to do any extra details or guard with us. He acted as though he were part of the Blood Angels group and that was that. If I were to be fair, it was probably because he was going on mid-tour leave soon and didn't want to get in a different rhythm, but I'm not fair, and this is another reason I took the man for $600 in spades games. This arrangement was convoluted to say the least, but I wasn't in charge and wasn't fully aware of the situation.

I had other fish to fry, and I was interested in feeling out Puffer and this new dick. I wanted badly to be the next team leader for our squad. Since Nance had left and SGT Zufall was quick to try and stay safe with the tanks, we only had SGT Crowley. He was and still is a good guy, but he was as dumb as two rocks put together. I say this with full respect for the man. I made his life hell. I questioned authority, I constantly took everything less seriously than it was, and I sometimes flat refused to do things. However, SGT Crowley treated us like a father and was there for anyone if they needed it.

He made sure we had hot coffee on guard and that everyone was taken care of. He watched the vehicle while we ate and tried to be as fair as possible if he had the choice of what detail we were on. Once he got so mad at my bullshit, tried to hit me in the face, and purposely missed. He cracked his knuckles on the window of our vehicle, which cracked as well. I realized this man almost fucked me up beyond belief and shut up. After I did what he asked, we started joking that he was trying to send me home early and would have nearly killed me. He began profusely apologizing and had an honest and sincere emotional response to the event. I kept telling him I didn't care and didn't hold any grudge. I knew I was a shithead and probably deserved it. He also had a moment when we were on top of a building assessing an IED. Another platoon was trying to cross a field on foot, and we were worried the other sniper brother might be using this to draw out soldiers. It was deemed a suicide mission and to call EOD. Davis and I popped up telling SSG Haas we wanted to go on a suicide mission, and that it was cool as fuck. SGT Zufall was hyping us up, telling us it was fine. Yet SGT Crowley got enraged and told him to go fuck off. He said to keep that shit out of our heads, that we were too valuable. I felt bad for SGT Crowley because he cared more about my life than I did. I was here to die in combat, not make it through.

The result of this mission was then to place us on what they called SKT (small kill team) and use our squad to assist snipers. We would get attached to a sniper team when they had intelligence about terrorists trying to implant IEDs, and then once the sniper team identified the target and shot them, we would lay down suppressing fire in the streets to clear areas and allow them to move safely back to a vehicle or our base. On a few occasions we were lightly engaged, but we never fought forward, as our only goal was to cover the snipers as they escaped from their hiding spots. I don't remember killing anyone, but there was a time when we finished our round of covering fire and rounded a corner to see the entire road covered in chickens. Davis and I just looked at each other and mowed down as many chickens as possible until a round hit the wall next to us. We turned the corner and ran back to the COP.

Finally the replacements began to arrive, and we started our rounds of vetting who was brand new, or had time in, or what kind of people they were. We got two replacements, Puffer and Valverde. Puffer had been deployed to Iraq before, got out and rejoined the Army. I made fun of him for coming back to this shithole and the Army, but I would later end up realizing how bad things were and do the exact same thing. Valverde was another thirty-year-old late bloomer. I knew Puffer would be the pick for team leader. He was, but was usually more involved with whatever was going on with SSG Haas and the other NCOs to take care of anyone. I began to show Valverde the ropes. We brought him out to the fuel point and burn pit and over to the tanks to help change track like we had done. We made the most of our time, to try and show this guy everything we could before we left Balad. One day while showing him the different weapons systems, we found out he didn't know how to clear an M9 pistol. I showed him how to clear it, perform a functions check, and take it apart. Every time he got it wrong, I smoked the dog shit out of him. We then went through all our weapons systems, ensuring he knew everything, because the last thing we were going to have occur was our new private shooting us in the back because he forgot how to clear his weapon. We made sure he knew that, once we left, shit was going to get real. It was going to be hell, and there was no real way to prepare. He just looked at us and smiled and told us he was ready. We headed out to the FOB with our new vehicles and fresh start. It was time to enter Hell, and the heat was rising.

ONE, TWO, THREE, FOUR.

YOU CAN'T COUNT TO FIVE.

ONE, TWO, THREE, FOUR.

CAUSE YOUR BRAIN IS FRIED...

ON CRACK! AND DOPE!

AND ALL THAT SHIT YOU SMOKE

CRACK BABY!

CHAPTER 17

Mordor

We approached the FOB during the night. There had been a few stops on the way, and we were convoying with CPT Austin, so he had other mission priorities than to just get back to the FOB. We were looking forward to a hot meal and a decent night's sleep. As we turned down Warpig the radio began to go off, and I asked Melsha what was going on. He told me lights were flashing from the FOB, and they were thinking there was a problem. I lit my cigarette and made sure everything was in place. SSG Haas was trying to call up on the radio to the gate guards but wasn't getting a reply. We increased our speed, and soon we could see red and green strobe lights flashing from the FOB. We were unsure as to what they were, but with the speed and erratic behavior it looked as though this was a distress call. I stood up and mounted on the gun, ready to begin pumping .50-caliber bullets into someone. We sped to the gate to see the guards casually standing around. CPT Austin jumped out of the Bradley, and we dismounted the HUMVEE in suit. He was asking what the hell was happening, and we heard the guard reply that it was salsa night, and the lights were from a makeshift club they had created in the gym. We stared at each other in disbelief. He looked back and asked the guy if he was fucking with him. The guy said no and he was about to go right after his shift. CPT Austin asked us if we would prefer to go to the COP for the night, and then come back the next day for our weekend reset. We agreed and returned the next day.

The next day the mood was completely different. When we rode up to the gate everyone looked depressed. You could barely hear the radio, and we saw no signs of any salsa night escapades. CPT Austin asked the gate guard how salsa night had been, and he said it sucked. The FOB had been attacked the night before, and mortars had been strategically fired at the lights from the gym. We laughed, and CPT Austin gave the guy a look of disbelief. I remarked that maybe next time they won't forget there's a war going on outside that wire, and the guy began to cry. I laughed harder, and SSG Haas told me to shut it. I stopped, but inside I was laughing that reality was just now making itself apparent to the window lickers. In my mind, they deserved everything they got. Most of the post was shut down, so our weekend was less eventful than usual. When we did try to go to the PX before we returned to the COP, the POGs were hyper fixating on the fact that we weren't wearing our protective gear and being cautious. We started using the bullhorn to tell the shits to keep their weapons at the low ready. Remember, POG, you're at war! We returned to the COP and spilled the tea from our trip.

We were in for a huge change of scenery. Apparently, the powers that be had decided while we were away that the Iraqi Army was supposed to occupy half of the COP, and now we were going to work together. We were supposed to take these guys on all our patrols, and each squad would be an extension of our own squad. The first few patrols we almost forgot the little fuckers, and once we went out it was apparent that we shouldn't have these guys with us at all. We began clearing palm groves, farmlands, and small buildings in Tahrir while trying to teach these idiots how to patrol. They were dirty, stupid, and useless. Our trash began to increase in quantity, and we forced them to build their own outhouses on the COP. We didn't trust them as far as we could spit, and our suspicions were confirmed when the first week some of the IA members tried sabotaging patrols because they had ties to Al Qaeda.

What came next was a surprise and did more to establish relations between the groups than anything. The general that oversaw the IA had locked these individuals in a cell on the other side of the COP. He would take them into a room and begin torturing and raping the men in there until they gave up their contacts. Within a couple of weeks, he weeded out the spies and gave them to us for processing. When we brought back prisoners for questioning, he would do the same. During one patrol we were shot at by a sixteen-year-old boy, and he was handed over to the IA with a sandbag over his head. He was beaten and raped until he gave up his contacts in his village. We were prevented from doing anything like this, especially with the incident at Abu Ghraib drawing light to some of the war crimes being committed by contractors and soldiers. However, it was their culture and their world. If they wanted to rape and torture one another, who were we to stop it? We had tried to get the kid to talk to us before sending him to the Iraqis. We had our interpreter tell him what was going to happen, and that it would be better to talk to us. But I wouldn't have believed me either.

This general had such a fascination with raping men that he even began to make kissing faces at some of us. Most of us would flick him off and he would laugh, but one soldier was especially attractive to him. It's sad, I can't even remember the guy's real name, he was a replacement. We called him Sweet Pea, because the general would blow him kisses, and we joked that he needed to go next door and score us a pack of cigarettes. This poor kid took the brunt of our mockery, and it had nothing to do with him being a bad soldier or any of that. He was just catching shit from people who had lost all sense of decency, morality, and humanity. If anything, if we had inspired him to go over there and beat that general to a bloody mess, we would have cheered him on and helped. What this also was a response to was the fact that we couldn't stop what was going on. We were told it wasn't our place, our morals weren't their morals, we had to let them figure out their own society. If this was the case, though, then why the fuck did we fly two thousand miles across the ocean to "help" these people? Their morals, society, and culture hadn't stopped them from committing genocide on the

Kurds, or improved their economy by spurring industrialization, or create any fucking Renaissance of art to inspire their souls. These were the same people that I witnessed hanging a gay man in the street because it's okay for a grown man to rape little boys, until the grown man gets married. These are the same people that I witnessed execute a woman by burying her and throwing rocks at her head until she died, because she was accused of cheating on her husband, who had three other wives. These were the same people that were selling their children into slavery and giving them grenades to give to soldiers in a town square.

The next addition to the suck fest that was becoming normal life was the addition of what our government called the Concerned Local Nationals. This name changed daily, and just like the terrorist organizations that existed in the region the name was always something about protection or freedom, and the men involved were just criminals and thugs. We had a list of offshoots of Al Qaeda groups in the area, and they were always something like the Warriors for Freedom, or Islamic Freedom Coalition, but they were terrorists trying to kill us and to impose a religious fascist state. The CLCs/CLNs (concerned local citizens or concerned local nationals) had gone through a systematic background check and processing to be a part of their group. Before this it was every man for himself, and often neighbors were settling disputes with RPGs. Now, we brought the city into the Girls College and put biometric data in a database for each citizen. We didn't require this, but we offered incentives to the people that participated. If we identified a criminal or low-level terrorist, their record would be wiped clean and they could begin seeking government services or participating in their democracy. They could get passports and other things like that, if they participated. This caused a swell of people to show up, and we ended up having to borrow two POG girls from another unit so we could check females at the gate and take their readings. We had conducted searches on females before, but only during battles. The men would act like you had just raped their wife because you checked to make sure she wasn't hiding a weapon or bomb, so we would end up having to beat the men to shut them up.

We input the populace in the system and sent it up top and finished the mission in a week. We had been away from females for so long that we were like dogs around them. We couldn't stop trying to talk to them or be on some detail with or near them. You could smell their scents around the corner, and most of us were salivating. Barton did the most flirting, but at one point or another we were all trying to fuck. Back home these women would not even have been on half of these guys' radar. They weren't especially ugly, but they weren't beautiful either. Yet here, they could have walked up to any of us and demanded we fuck right there in the open, and not one of us would have said no.

The men who then wanted to sign up for the CLC group were identified and given their instructions. Like most things officers and politicians invent, this was a good idea. The gist was that local men would patrol their own streets during the day and night, and that would deter the terrorists from operating in these towns. People would feel safe, bring us more intelligence, and foster better relationships. But, since the good idea fairy had bestowed this idea to officers, no one had thought about the effects or challenges that would arise. All the men who signed up for biometrics and CLC duty were felon-level criminals and low-level terrorists. These were the same assholes who had killed our brothers just a month prior. Now, not only were we supposed to look these fucks in the eye and give them a paycheck, which was larger than we made a month, but we also were supposed to stand by as they raided their own towns, raped women, murdered men, and forced the citizens of their neighborhoods to give them protection money. At one point the army, the police, and the CLCs got into a firefight with one another. We were called out to respond, and the CLCs had killed most of the IA and IP. They stole their machine guns and other weapons and went back to their compound to dish out their loot. We had to go over and take back the machine guns, telling them they were only authorized to have the AK. As for the murder of the soldiers and police, we didn't do a damn thing. The CLC program was the brainchild of President George W. Bush. This fuckface thought that by giving previous illegal militias, terrorists, and criminals weapons and authority, they would in turn help us fight Al Qaeda. So, if President Bush gets around to reading this, know this small part is for you. Your idea literally resulted in the mass rape of women and boys, societal anarchy, criminal extortion, and murder. These things were happening before you had your dumbass idea, but when you gave the CLCs guns and authority, you ensured that it would happen and be supported by the state.

The heat was rising, and so were tensions between the local population and new groups. We would be driving through the town and watch these idiots in their orange vests flag us their weapons. Most of the time I would stand up and aim straight at them, and they would drop their weapon. We would shout at them in Arabic, or have the interpreter speak to them. When Snake talked to them, he would slap them in the face and yell. If they even tried to act like they would fight back, we would surround him and point our weapons in their faces. We wanted to kill every single one of these assholes, and the fact that we had to play nice ate every man inside. However, that didn't stop us from finding any reason to kill them. We had an idiot in a roundabout trying to direct traffic with his AK in one hand. He shot the rifle and rounds rang out close to our vehicles. We all turned our guns on him and riddled him with bullets. We stopped, grabbed his AK, and left him there for his CLC buddies to pick up and maybe learn a lesson. However, they didn't learn. We had multiple other engagements while working with IA members, and these gangs of idiots would try to ambush them. We usually ended up mowing them down and taking their shit.

At one point a gunner from another PLT ended up getting into such a heated argument that he killed one of them, and there was UCMJ involved. He lost his rank and had some extra duty, but no one was going to go to jail for killing the trash that had been trying to kill us this entire war.

This ended up as predicted, not working. Tensions, crime, and terrorism were higher than ever. Other units were losing people weekly, and even the 5-20 BN lost an entire Stryker, and their crew was wounded by rolling over an IED. At one point it was time to do some more raids, but before that could happen, we had to get briefed, for the first time, by PSYOPS.

It seemed like a whole platoon of these guys came to our COP and were running around doing God knows what. By the end of the week, we finally received a brief on what we were going to have to accomplish. The Surge, as we were coming to call this event, now needed to ramp up. Everyone would be involved in missions, patrols, and operations. We were going to start implementing curfew. This was going to be a policy of martial law that made it illegal to be outside after dark. We were on orders to shoot anyone after dark once curfew was in effect.

Before this policy was implemented, we needed one week to move around with the PSYOPS and Civil Affairs crews to inform the local populace about the changes. We went street by street, house by house, to pass out flyers and have our interpreters tell each person we could about the changes. I pulled security from the vehicle for the most part, but when other guys got tired, we would trade guns and I would walk streets. I went with Snake and we came across a man playing what looked like Slapjack with some older gentlemen. We told them of the changes, and the old men blew us off, waving their hands. Then the first man began speaking broken English. "Mista, me love America! America so good! George Bush King, king of Dick Sauce!" Dick Sauce was how the locals pronounced Texas. It always made me laugh to hear that Bush was King of Dick Sauce, and I wasn't even politically interested at the time. I kept trying to calm the man down and tell him that if he was out after curfew he was going to be killed, but he kept repeating himself and grabbing my sleeve. I told Snake to tell him once more, and he told me he already had. We gave him the flyer and moved on as he kept shouting. I looked at Snake and told him we were going to shoot that man next week, and he laughed.

Curfew was implemented, and the first week we saw sporadic activity. After a few farmers and other idiots were shot at after dark, it seemed like the message had gotten through. To our knowledge at the time, no one had been killed over curfew. That changed one night when we were on a walking patrol in Tahrir. We were showing the IA how to use night vision when a man appeared in the middle of our patrol. PVT Rivera just let him go by, and he stood in the middle of our formation. Then he began walking to the back, towards me. I shouted out to the front what the fuck was going on, and everyone turned to hush me. However, SSG Fili saw the man and pointed. He turned and it looked like he had a stick or something under his robe. I brought up my SAW and fired. As he lay in the street screaming, SSG Fili went up to the man and kicked something away. An AK appeared under his robe, and SSG Fili shot him. When I moved up to where the body was, I could see the card player looking straight back at me. Snake happened to be with us, and I heard him say, "See, these Arab fucks are retarded." We called it up and kept moving. I couldn't help but think we were going to have hell to pay for this.

AND NOW THE BATTLE'S OVER
AND THE SMOKE IS ALL AROUND
WE WANTED TO GO HOME
BUT WE'RE SIX FEET UNDERGROUND!

CHAPTER 18

Death from Above

There was an Iraqi Army lieutenant who was a complete waste of life. We had been patrolling regularly with the IA units, and they were now forming into more official squads with leadership. However, they chose their leadership by who had the largest mustache. So this one lieutenant, who had an exceptionally large mustache and Teddy Roosevelt-style glasses, was already giving me weird vibes before we even started what was supposed to be a normal patrol. He was extremely small and thin but had an arrogance about him that seemed undeserved. He had these long boots on and looked more like someone about to go on a hunt in the bush than combat. As we lined up and began getting ready, the IA squad was placed behind my team. I was told the interpreter, Snake, would be able to come back any time I needed him, and to just ensure the squad kept their spacing and did not shoot us in the back.

As we stepped off, the LT began holding his AK behind his back like a whip and goosestepping and walking all around the formation in a random fashion. He was also flagging our squad with his rifle and paying no attention to anything that was going on. I could not believe what I was seeing. I brought our team to a halt and called Snake. When I asked why the LT was acting like that, he said he was in charge and did not have to act like everyone else. I told Snake to tell him he needed to assume a security posture and be an example to his men. We were in a warzone, and he needed to act accordingly. He looked pissed but just nodded. As we began stepping off, he did not goosestep, but he was constantly walking in front of his men and everyone else. I once again halted the formation and got Snake to relay that if he did not stop this bullshit, I was prepared to strike him. He shrugged his shoulders, and we began moving again. Before we could even get down the street, he was back at it again. I waited until he was completely looking up in the air and no longer paying attention to butt stroke him in the face. He hit the ground, bleeding profusely, crying and wailing, then turned around and ran back to the COP. When we finished the patrol, I saw my CO standing at the gate looking like a disappointed mother. I told him what had happened, and he agreed that I had done the right thing. They spoke to the Iraqi Army, and we never saw the guy again.

We continued patrolling through palm groves, and all of us were so hungry we started grabbing the fruit from the bushes. When we got to the end of a row we would move forward and on to the next row of fruit. SSG Haas started to call everyone out for not pulling security or scanning, until he came out of his row, and a grape rolled down his IBA. We then started pulling foot patrols daily, and each squad leader led at least one per day. But since this consisted of a night and day patrol, most of us ended up patrolling at least twice a day. My thought was to try and get out of the COP as much as possible, so I would volunteer to go on patrol whenever I could. This time I went with SSG Fili, and the small patrol ended up turning into a full-blown mission in Tahrir. SSG Fili turned back towards me and told me to keep the IA at a distance. We were moving forward in Tahrir but were turning up much of nothing in our search. He told me we were about to have the Air Force conduct strafing runs to see if we could stir up activity. As men began to move through the street, we could hear other platoons returning fire across town. We moved around our building, and I kept turning to ensure the IA were following at a safe distance. Apparently, curfew was the last straw and those who wanted to fight about it were ready to play. We cleared through our portion and began seeing the jets swoop in above our heads. The chatter was picking up on the radio as well as the dust from the streets. SSG Fili told us to take the Velcro off our KPODs so the Raptors would see the IR squares and be able to identify friendlies. He said they were about to drop a JDAM, and we needed to get out of the blast zone. We exited the gate of the house we were using for cover and turned the next street over to see a Raptor in full glory barreling right towards us. We cheered it on as the engine roared, then looked at the end of the street to see a wall of dust flying at us. SSG Fili yelled for everyone to take cover. I felt something like an IED, but bigger, fill the void around me. I was knocked over by what felt like a wall of wind, and rubble and dust flew at me as I hit the ground. I couldn't hear anything, and the street was hazy. I couldn't tell if it was the dust, or if I was cross eyed, but now I was wondering what had happened.

As my vision focused, I sat up to see my SAW still around my waist and the other members of the IA and our squad covered in rubble. Everyone was moaning in pain, and SSG Fili was screaming on the radio. At first I couldn't make it out, but then it became clearer, and I could hear him announcing, "They just JDAM'd us!" These fucking idiots had dropped a 500-pound bomb on our heads and torn up most of the street. The air was hot, and rubble was still falling, when he looked at me and screamed that we had to move. Another run was coming. We began to run down the street as fast as we could. That's the only time I remember not caring about anyone's life other than my own. I ran, and since I was at the front of the formation, I was outpacing everyone, until I heard another bomb drop. The wall of wind hit my back, and dust and rubble flew forward. I kept running and made it out of the dust just well enough to see another squad. We had been at the very edge of the blast radius and had just made it out. Doc McCollough came over to check us out, and we were all fine.

We then moved in closer to the city just as the return fire was intensifying. Doc was with SSG Fili and was trying to move back to 1SG location. As he argued with SSG Fili in the street, I looked to my right and saw a man lifting an AK and taking aim at Doc.

I lifted my SAW, and in slow motion I watched as the rounds led straight up his body from his feet to his head. As he fell back, I saw another man with an AK trying to step over him, and I held down the trigger until the belt ran out. I then saw two other men fall over. Doc looked surprised and said we needed to go check them out. As we all moved forward and I changed out my nut sack, we looked down on them. They all wore black and had white Arabic lettering on their chests. Doc commented that one was moving, and I let out another three bursts on his body. SSG Fili called up BDA and we moved on. Then out of nowhere a giant bomb went off in the middle of the city. It looked and sounded like the first bomb on my left seat/right seat ride. I could see a plume of smoke rising from the buildings in front of us, then SSG Fili's radio started crackling and going off. I could hear people screaming and then battle roster numbers. SSG Fili informed us that three of ours had just died.

SSG Shaw, Camcho, and Romero were guys from second platoon. They had been clearing the other side of the town as we all moved in. At first I thought it must have been an IED because Romero was a Bradley driver, but SSG Fili told us it was a building that was rigged. Apparently, there was a rat nest – a bagged group of grenades attached to a wire that pulls a pin when you open the door – in the middle of the room they were trying to clear, and the explosion was so big that it brought the building down on top of them. We were instructed to move to the perimeter and form a cordon. The tanks and Bradleys showed up to help with the clearing of rubble, and it was late into the evening before we got back to the COP. Next thing I remember is I was once again holding an M16 in my hands and firing into the sky while listening to people who never knew these men act like they were their closest family members. I felt numb. I didn't give a shit, and the only thing that bothered me now was that fewer people would be available for details and guard. We would be asked to cover down on their COP shifts because they didn't have enough guys now.

We got our weekend at the COP, and I was filling up on cigarettes, dip, and candy when I overheard some talk about mid-tour leave. I had concluded that I wasn't going to get mid-tour leave because 1SG had told me as much at the beginning of our deployment. I understood that it was a punishment for my Article 15. I would always pipe up and tell people they should feel lucky because I wasn't even allowed leave. This was already a long deployment, as the Army had just implemented one-year deployments the year prior. Most other units had done six to nine months, and most guys with multiple deployments would end up with the same deployment time as we would by the end of this deployment. Soon after, we were getting called to the FOB for a briefing. For the first time since October, our whole company was together on the FOB. There were a small contingent of forces left at the COP and embassy, but mostly we were all here. 1SG and CO informed us that everyone was getting a three-day pass to Kuwait. Apparently, our battalion had been investigated by CID for sending up the wrong numbers of troops and found some sort of fuckery attached to the situation. We were under strength and had been so low on manpower that, by Army regulations, we were supposed to be pulled back to the rear and refitted. Yet due to this misinformation, we had been fighting under strength. However, to balance the news, we were also being extended on our deployment for three months. Those like myself that were on ADVON (first arrivals into country) were not supposed to be on REARVON (last leaving country) but, as luck would have it, my name came up on that as well. This means everyone else would have fifteen months deployed, and I would have two extra months. I lost my shit and began cursing. SSG Haas said he would take care of it, and we all got ready to go back to the COP and embassy.

Then, with no warning, a bald lieutenant stepped out from the CO's Chu and we were introduced to our new platoon leader, LT PJ. Just a few weeks prior, LT Sparks had been injured with the 5-20 guys in a Stryker. They ran over an IED and the blast was so large that it sent him flying outside the vehicle. I don't know what happened, or if that caused him to decide to move or he was forced to move, but he was relocated to the S Shop. At this point we had been losing LTs like a bad habit, and it was extremely unlucky to be our platoon leader. When LT PJ arrived, he had some shit to say, before SSG Haas shut him up. He told him that if he didn't want to end up like the last four fuckers, he would shut the fuck up and just watch what went on. He told him not to worry about his end of campaign medal, or what went down on his ERB for his career, and that he would be taken care of. However, if he couldn't learn the lesson of the last four guys, then he was destined to end up injured or worse. Things seemed okay from our point of view. The LT and the PSG weren't fighting, and he was acting more as a figurehead that would give out mission briefs. Yet as time went on, he let more and more go to his head and started trying to take the reins from SFC Reynolds and SFC Malo. Now each brief had some tension to it, and LT PJ decided he wanted to go on these foot patrols all the time. SFC Reynolds didn't go on the foot patrols, and neither did 1SG. They planned missions and kept supplies running. They made sure the company worked, and it seemed like maybe he was a bit jealous of the authority these NCOs held. In truth we were destined to be without a LT and, like his predecessors, he would have to learn the hard way.

I'M SITTING IN MY FOXHOLE
THINKING ABOUT LIFE...
UP POPPED HAJI!
I STABBED HIM WITH MY KNIFE!

CHAPTER 19

THE TEST, PART 2

Bullets skipped off the pavement and began traveling down the road towards my feet. The IA huddled against their sliver of a wall and were screaming. I began to scream at them to pick up their rifles and shoot, as I continued to let my SAW use up another belt of ammo. I screamed at SSG Fili that I needed to reload, and he got Davis to fire an M203 at the building. While I was in the open trying to explain this guy's location to Davis, more rounds started hurling down the street. I could hear some cries ahead from the IA members, and I grabbed one guy and told him they were all going to die if they couldn't get the gate open. As one unit, they began smashing the gate in a rhythmic slam with their shoulders. Eventually they were able to pry it open enough to pull the latch and break inside. Before I entered, I stood in the street and dumped the rest of my rounds. I saw the rounds getting close to Davis, so I adjusted my aim to where I saw the flashes coming from, then watched as sparks flicked off the pavement and moved my way. As the belt ran out, I noticed the IA squad had finally busted through the gate, and I made my way inside.

For some reason I was completely out of breath, and I could barely stand. I took a few seconds to compose myself, then stared at the motley group in front of me. Some of them were in the corner smoking already, while one man was laid out against the wall crying and another was on top of him. I ran over and saw he was bleeding from his femoral artery. I took out my tourniquet and wrapped it around his leg. As I turned it, I told him this was going to hurt, but I'm sure nothing could have prepared him for the pain, even if he could speak better English. He began squirming and screaming. I pushed him back down. The whole time my SAW lay flat in my lap. I kept turning until the bleeding stopped, then used his blood to draw a T on his forehead and the time. Another IA member tried touching it and I slapped him and said "No!" as loud as I could. They began praying *Allah-u-Akbar* and I loudly exclaimed, "God isn't going to save you today. You either save yourselves or die." I went back to the gate as I changed my belt again and began firing. I was feeling smart about bringing my extra rounds now, as no one was able to hit this guy. I could see SSG Fili, Davis, and interpreter Steve in a courtyard opposite mine. I then looked down to see that my leg was completely red. I thought, How did that guy's blood get down there? until I realized I had been shot.

I then reached for my IFAC, and it was gone. It had been ripped from my side and was in the street in front of me. I went back to the IA and asked if any of them had a bandage, and they all said no and started hiding their bandages. I spat in disgust towards them and moved back to the gate. As I approached, I yelled for SSG Fili and Davis to provide cover and ran back into the open street to get my IFAC. While running back I could see Rivera laid out against a gate in front of SSG Fili. I ordered him to fire back, and he began fumbling with his gear. I could feel the bullets at my heels, but I made it in the gate and wrapped my leg. I walked back out and began shooting. I was screaming at Rivera to return fire. He told me he had been shot, and I replied that we all had been shot.

At that point a horror beyond horrors occurred. He pulled out of his front pocket a grenade. Now flash back to the COP before we headed out, and PVT Rivera was with Davis and me in the ammo closet. He was begging for a grenade, and I kept telling him no. I told him I had been in the country this whole time and never blown anyone up with a grenade. I'd be damned if a private got a grenade kill before me. He laughed, and Davis gave him one. Now this little fucker was prepping a grenade for some reason. Was he going to kill himself? I couldn't see any other enemy fighters, and when I looked out of the gate down the street, I saw SGT Crowley face down in the street. That's when I noticed barrels peeking out around the corner from his position, and before I could make out what kind they were the grenade went off. I looked over at Rivera, and he had thrown the grenade behind himself and was using his body to hold the gate closed. He said he had heard men in that courtyard, and they had tried to open the gate, so he threw the grenade. I then looked ahead to see another four men in black turning the corner. I fired my SAW and dropped four of them. This caused the machine gun nest to target my wall again, and I jumped back as the debris flew towards my face. I looked at Rivera to scream for him to return fire again, and he weakly lifted his SAW and let out a small burst. I let my belt go and realized I only had 200 rounds left. I had gone through 1200 rounds, and my last drum was all that stood between me and any more enemy fighters. I walked back to the IA, sat down on a cinderblock looking at the center of the gate, propped my leg up on the block, and aimed my SAW at the gate. I pulled out my pack of cigarettes and there was my last smoke, my lucky turned upside down. I chuckled and said out loud that it was a good day to die.

The gate then burst open! Two American soldiers kicked it in and came running at all of us with their M4s raised but did not announce their entry, and I was ready for the worst. I had lost so much blood by this time that when I fired my SAW, I missed every round. The guy in front dropped his weapon and his team leader came in screaming and pointing his weapon at me. I screamed back, "No one announced friendlies coming through!" He started kicking the private and then told me he was from 5-20 and they were here to help. He escorted me out and by now the firing had stopped. I looked up and saw the Blood Angels tank and the TC yelling at Fili about where the enemy was. He couldn't tell him, so I ran to the tank and told him third floor, middle window. He sent a main gun round into the window and knocked out most of the building. I began walking back to the Stryker at the end of the street, as the others took on the other wounded. They were moving Rivera to the Stryker and asked if I needed help. I told them I was okay as I limped forward. Davis came to my side and began bracing me, and then the interpreter Steve also came to help. We got to the Stryker and found out that Doc Nguyen had been killed and SGT Crowley was dead. I let out a wail and turned away from the back door. Then someone piped up saying he was alive and on his way. I asked Davis for a smoke, and he gave me one, and we helped load up Rivera, SGT Crowley, and the IA who'd had his leg shot. Then LT PJ was being carried on a stretcher to the Stryker.

I looked at Davis and he told me that during the fight Doc had pushed LT PJ inside a courtyard, and when he looked back outside, he took a round to the neck. LT PJ tried to put pressure on his neck, but he bled out and died. LT PJ had taken rounds under his vest and had been shot in the ribs. Then, as we all were loading up, the Stryker team tried to push me out, saying only wounded would get a ride. I told them I also got fucking shot and pointed to the Israeli Wrap on my lower leg. The guy scoffed but let me ride. We held Crowley's hand while the Doc began working on him as he was writhing in pain, and every bump was excruciating. He was trying to laugh and joke, but then a pothole would jolt him back into a state of pain. I told him his ass was too stupid to die and he was going to make it. The medic did get a line started, and pushed some pain meds, but mostly he was trying to stop his bleeding and locate all his wounds. He had been shot in multiple locations, and eventually it was ascertained that he had been shot nine times. We would joke with him later that he was now just as hood as 50 Cent, but during this moment it didn't look good for him. The Stryker slowed down and was being stopped at the gate and the gunner freaked out on the guards. This was FOB Gabe, and it was commonly used as a CASEVAC (casualty evacuation) point while we were in Baqubah and Tahrir, and they knew when a medical vehicle was coming through to open the gate. The gunner would radio up a special code that changed daily and that would alert the gate guard to let the CASEVAC vehicle through. I heard the gunner cuss out the guard and throw a piss bottle at the kid, and I started laughing. They began bringing the others inside, and I walked into the CASEVAC station. The PA (physician's assistant) CPT Roach began working on the others and after a minute came over to me. He was the PA from back at Fort Hood who was a complete piece of shit. He constantly accused soldiers of malingering and would often refuse to treat people. Everyone hated this asshole. At one point he had told me I couldn't be allergic to Benadryl (I was) and that I was an idiot who imagined having anaphylaxis at age six after taking it and that my grandmother was also allergic to it. I called his bluff and asked him to give me some, and he said he couldn't because it was on my records, but that he still didn't believe me.

I had been waiting for the opportunity to revisit this issue. As he was asking pertinent health information he got to my allergies, and I said, "I'm still fucking allergic to Benadryl." He scowled at me as it suddenly dawned on him who I was, and he quickly left to go work on my teammates. I told him I was shot in the leg, but stable. He had a medic start an IV on me and then they let me go smoke, so I walked outside and met Davis. I told him about Crowley and Rivera, and what I had said to Roach, and he laughed his ass off. Just then we could hear them screaming inside, and I commented that now that everyone was out, they would be pulling double guard shift, and that was bullshit. He laughed and said I should worry about myself, but I said I'd be back soon and not to worry. As they got us prepped to leave for Balad in Baghdad they tried using a wheelchair to take me to the helicopter, but I protested and said I wasn't a bitch without legs. As I got up, I lost all feeling in my leg and fell. They scooped me up and sat me back down and loaded me on the bird. I thought I looked cool as fuck, but the guy who helped probably thought I was an asshole. I sighed to myself as I lay down on the cot and thought, "This is it, you're finally worth something."

MY GIRL'S A VEGETABLE SHE LIVES IN A HOSPITAL
BUT I WOULD DO ANYTHING TO KEEP HER ALIVE
SHE HAS A GREEN TV IT'S CALLED AN EKG
BUT I WOULD DO ANYTHING TO KEEP HER ALIVE
SOMETIMES I PLAY A JOKE PULL THE PLUG AND WATCH HER CHOKE
BUT I WOULD DO ANYTHING TO KEEP HER ALIVE
SHE MIGHT NOT LIVE THE NIGHT SO SHE WON'T PUT UP A FIGHT
BUT I WOULD DO ANYTHING TO KEEP HER ALIVE

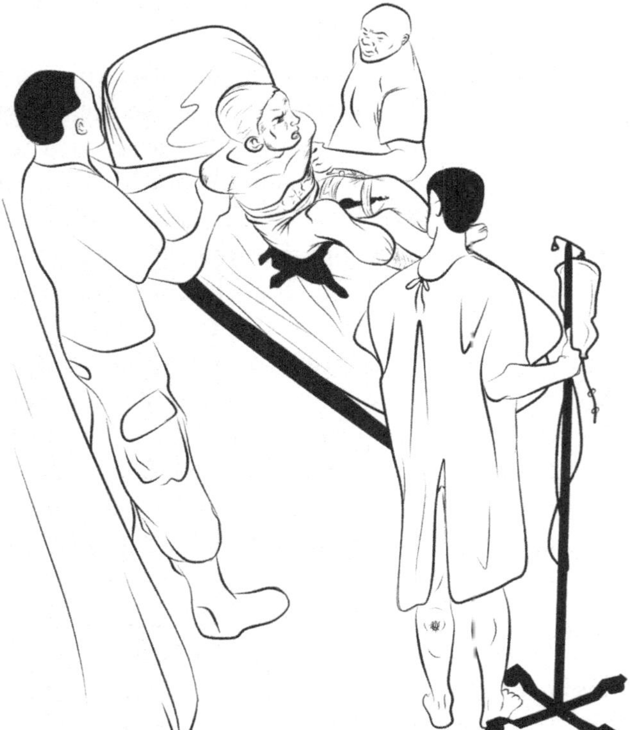

CHAPTER 20

Don't Rest upon Your Laurels

The helicopter landed, and a team of people were waiting at the flightline for us. There was a chaotic whirl of movement and noise, and I couldn't follow what was happening around me. Before I could figure out where I was, I was being placed on a gurney and a blanket was laid over me. A nurse told me I would be prepped for surgery soon and to relax. I tried to doze off, but the commotion was too much, and I lazily looked around the room at the nurses moving back and forth. I couldn't believe this was a fully functional hospital inside a tent, but everything was like any other ER back in the States. I was given some water by a tech who then informed me he had to look at my wound. First, I'd like to explain that an Israeli bandage is meant to be placed on a wound and pressure wrapped to control bleeding. Due to how it's made, you can easily unwrap the bandage and rewrap it if necessary. Yet this tech thought it was a good idea to use his trauma scissors to cut my bandage, at which point he exclaimed it was a gunshot wound. I told him I could have told him that and we both laughed, but as he began to run off to his next patient, I asked him to please get another bandage. He assured me I was going into surgery soon and it would be fine. Yet hours went by, and another nurse came to assess me. I told her why I was there and that I needed my bandage reapplied, but she said the same thing the tech had said. She also looked at me funny when she asked what time I had been shot and I told her it was only a couple of hours ago. She asked if it had been daylight when I was shot, and I said yes, and she informed me it was almost midnight. I was startled that my concept of time was so off, but I guessed I would be in surgery soon and it didn't matter.

At this point I told them I was feeling cold, even though it was at least 120 degrees outside, and she gave me two blankets. Then another nurse came by and noticed I was shivering uncontrollably. It felt like I was naked, outside, in the winter, and the look on this nurse's face was visibly concerned. I was getting drowsy, and when she asked how long I had been there I could only say the word daylight, and that they still hadn't given me a new bandage. She pulled back the blankets to see the entire gurney soaked in blood. She screamed and flagged down a tech for help. I began laughing and, as she said they were taking me to surgery, I replied, "In a few minutes, huh?" She looked startled at my reply, and I began to laugh. They rushed me down the halls, and I looked to the side and could see these POGs in the hallway. They looked bored, or upset, and it caused me to laugh harder. As they put the oxygen mask on my face, I locked eyes with another POG and exclaimed, "Al Qaeda couldn't kill me outside the wire, but these assholes are going to kill me here," and continued laughing. A man in a white coat and glasses rushed to my side and got my attention. He told me everything was okay, not to worry. I repeated, "Because I'll be in surgery in a few minutes," and laughed. He smiled and said, "No, you're on your way, son! I just want to know if we find a bullet in you, do you want to keep it?" I told him I would love to, and he began adjusting things around me. He then said, "Don't worry, you'll be going to sleep in ten, nine," and then darkness enveloped me. Everything went black, and nothing existed.

People postulate on what being dead is like. There's darkness and a bright light, maybe you're floating, then some great inner peace comes over you. Or if you were a terrible person, you might experience fear and some form of hell. Others see themselves on the table as they leave their body and find some fascinating reason to keep living, so they're sped back to their body to pursue some great life goal. For myself there was darkness, nothing, and then I woke up. Nothing, no out-of-body floating, no magical experience, no hell. It was just like another day, except now my throat hurt and I was barely strong enough to lift my head. I looked around to see rows of beds, and nurses working on other patients who seemed far more fucked up than me. I shook off the dizziness and rubbed my eyes, and the IV tube tickled my nose a bit. That's when I noticed a nurse at the edge of my bed, and she was smiling. She introduced herself as CPT Hepler, and my mind felt something familiar about that name, but I couldn't place my finger on it. I was trying to speak. She handed me a glass of water and said the anesthesia was wearing off, but she was CPL Hepler's wife. Then I remembered: He was a guy from second platoon. I laughed and looked down to see I was in some nasty green gown, and I finally coughed. She asked if I needed anything, and I said a pack of cigarettes. She smiled at me, then told me to get some rest for the time being. As she began to walk away, I managed to say, "Crowley." She turned and told me they were fine. Rivera, Crowley, and the IA soldier had been placed in the ICU but would be shipped off to Germany within a few days, and I would be going with them. I heard myself saying "No" as I began to doze off, then chuckled again. I stopped her one last time.

"Hepler! I remember him, he's a corporal. How are you guys married? I mean, I thought?"

"We were married before I commissioned."

"Oh, cool. Uh, did you see my dick?"

She laughed and said yes, she had seen my dick. I laughed back and said I couldn't wait to tell CPL Hepler his wife saw my dick. I then dozed off and blacked out. The next time I woke up I saw a group of people around me. There was SSG Fili, 1SG, and the brigade and battalion sergeant major, as well as the brigade commander, Col. Sutherland. He thanked me for my service and gave me a Purple Heart. As he left the others began to question me on my version of events. Some looked perplexed as well as entertained, and I honestly can't remember my account due to the number of drugs that were in my system.

When I awoke again, I found a carton of cigarettes, a soda, and four *Playboy* magazines on the table near my bed. I looked up and saw CPT Hepler and thanked her profusely. I realized at this point I didn't have clothes or a lighter and asked her about that. She said she would hook me up, and that physical therapy would be by soon to help me out of bed. I sat up and for the first time realized that my left leg felt heavy. I looked down and saw my calf was huge, like the leg of a Scottish Highland warrior. Then I saw the bandage, and another nurse was already on her way to check up on me. They removed the bandage and I saw staples and stitches, but it looked like something out of a horror movie. I didn't even register it as my leg. They then realized the staples and stitches were botched and said they needed to redo the process. They began to clamor for local anesthetic, but I told them to forget it. We just needed to get it done, and I was ready to go smoke. They stapled me back up and rewrapped my bandage, and before I could blink the physical therapist was at my side. She said they needed to fit me for crutches, and that I should start walking. I could blame it on the meds, but I didn't want to move, and I was enjoying being pampered, so I protested a bit. I couldn't believe I was already being required to start walking or doing physical therapy. I bitched about it for five minutes, then got out of bed and jumped on the crutches. They dug into my armpits, and I was surprised by the level of pain and the strength it took to hold my body up. Yet the therapist made a deal with me: If I could make it to the door, she would let me have a cigarette.

I crip walked my way down the hall, and I could feel my butt hanging out of my gown. I made it to the door and went outside to see the sun for the first time in two days. There was a group of people sitting around a table on a porch. One guy was getting a blood transfusion and had the box with blood sitting on the floor, and other guys were in flight suits. I sat down and lit my cigarette and began to talk. We traded stories where we had been, and I found out the transfusion guy was a marine. I probably went through the whole pack of smokes before the therapist came outside to get me. As she did the indirect alert sounded, and she closed the door. I looked at the other guys and laughed, and the soldiers made their way to the bunker, but the marine and I stayed where we were. I could see some woman under the bunker screaming at us and motioning us to come over. I kept smoking and laughing, the marine remarked about what bitches they were.

When all clear was sounded the woman furiously came over and began telling us that she was going to talk to the commander about how we disobeyed an order from a senior NCO. That's when I had enough and told her I would love to speak to the commander and relay the fact that they had decided to abandon two injured soldiers at the door for their own safety, how the man next to me was getting a blood transfusion and I couldn't walk without aid, but not one of her cowardly soldiers had offered aid, and she had refused to be a leader and put our lives above her own. She snarled and walked inside as I kept calling to her, saying I'd love to speak to her commander. The marine looked at me wide-eyed and laughed. I shrugged my shoulders. I wasn't going to suffer the wrath of some coward who sat back on a FOB while real men went out to fight. At this point I was still an E1 and had nothing to lose. I moved back to my bed, and on the way stopped to say hi to SGT Crowley and Rivera. Still doped out of their minds, they were being prepped to leave for Germany. Their case worker came by and said she was looking for me as well. I had been put on the manifest to go back to Germany and then the States. I told her I wasn't going. She looked at me and tried to repeat that I was on the manifest, and I told her I wasn't going to leave Iraq. SGT Crowley looked at me with confused eyes, and I repeated myself. I told them I would be okay in a few days, and I needed to get back to the line. There were now five fewer people to pull guard and details, and with SGT Crowley gone there were fewer NCO-led details and shifts too. I couldn't go back when I had only been shot in the leg and was going to be okay. This was even heavier on my mind, as I knew I wasn't even going to get mid-tour leave. I could go back to Germany at this point, and it would all be over. I could escape this hell and get one over on 1SG and everyone else. At this point, I could fuck the whole company and put the pain I had felt for so long on these assholes. But something told me no. I couldn't do it. I felt like I needed to be there for everyone, and my time wasn't up yet. I had just gone through the trial by fire, and now was I going to rest on my laurels? I hugged SGT Crowley and told him he was like a dad to me, then I hugged Rivera and told him he was an old bitch and to get well soon.

I spent the next four days walking as much as my leg would take. I had a wounded warrior shirt and shorts, and by the fifth day they moved me to the ward to prepare to leave. I hadn't brushed my teeth or showered, so I asked to do some hygiene and asked when I could go back home. The orderly looked at me weird, and I said I meant to Baqubah, that was home now. Wherever they went I went, and wherever you hang your hat is home. He confusingly said my chain of command was arriving soon and that I would be able to go back if there was room on the bird. I waited until dark and received another surprise. The IA soldier I had used my torniquet on had survived and had specifically asked for me. He spoke broken English better than I thought, and he thanked me and said I had saved his life. I told him it was nothing, but then he asked why I hadn't gone. I told him I had a job to finish and was fine. I was now just using a cane to walk, and he saw my bandage and pointed at it. I told him it was nothing and I'd be fine. He laughed and, as I began to leave, a group of orderlies came in to move him out of the way. I asked if I could help, and someone shouted to move.

As they did, I could hear screaming down the hall. It was the CO for 5/20, the same commander I would learn had ordered my rescue, now on the table being stripped and worked on. He had been hit by an IED and was dying. I don't know if he made it or not, but the screams I heard that day were haunting. He was in bad shape, and I had heard techs and orderlies complaining that they couldn't see him immediately because they were on orders to treat local nationals before us soldiers. My mind burned with rage, and I couldn't understand why they were just going to let this man bleed out and die so that a few Iraqi civilians could get checked out. It made less sense when I finally came home and saw the death toll of civilians that was allowed to happen anyway. I screamed that he was dying and fuck those civilians, and others joined in. They cleared an area for him, since apparently a mob was forming now, and moved the CO into ICU. I sat back down and rubbed my head. I needed to go, this place was foul and horrible for morale. I went to the desk to find out my sergeant major and battalion commander had arrived, along with the 5/20 sergeant major. I asked them if I could leave, but no one wanted to hear from a private now. I waited a minute then packed my bag and got my things ready. I asked SGM Harris if I could leave, and he said it was up to the doctor. I told him the doctor needed him to release me back to the unit, and he tried to circle back and say the doctor would decide. I grabbed him by the IBA and told him I was fine, I didn't need to spend another day in this purgatory. I was fucking going insane and needed to leave. Before he could lose his shit, the 5/20 sergeant major came over and got between us. He confirmed I could leave with them and would be dropped off at Warhorse. He then asked where my Kevlar and weapon were, and my heart sank. I didn't have either, didn't even know where my uniform was. He laughed and said don't worry about it, I could use his, and with that we were on a bird back to Warhorse.

As I got off the helicopter some POG tried to tell me to put on a cover, and I gave him a middle finger as I walked off. The SGM asked if I knew where to go, and I thanked him and said I knew where BCO was. I began walking back to the TOC in short sleeves, shorts, and flipflops. I must have been stopped a dozen times, and each time had to explain I was back from medevac and trying to report. Some

NCOs just told me to go, some were still trying to question why I didn't have a uniform or PT belt, or where this and that were at. I just yelled I had been shot five days ago and needed to go to my company. If they weren't going to give me a ride or help, they needed to get out of my way. I finally reached the company TOC and walked up the porch to our building and opened the door.

1, 2, 3 AND A QUARTER

I GOT A DATE WITH THE GENERAL'S DAUGHTER

GENERAL GENERAL HE'S A REAL SUCKER

HE DON'T KNOW THAT I'M GONNA FUCK HER!

CHAPTER 21

Back in the Saddle

I sat over a barrel watching the flames dance like ballerinas in *The Nutcracker* while inhaling the black smoke blowing in my face. A soldier named Larson had become the supply sergeant after we had moved our previous sergeant out. He had my kit and was burning anything with blood on it. He then took out the uniforms and gear from the other soldiers who had been with me in Tahrir. I wanted to take Doc's nametape off his IBA, but it was covered in blood, and we had to burn that as well. I watched as the last piece of him was engulfed by the flames and made a joke about worrying I had contracted AIDS from the IA soldier to break the uncomfortable silence. He then gave me a new issue of uniforms and gear and left me to the barrel, asking that I put it out when it was fully burned. I sat there smoking and wondered what would happen next and when I could get back to the line. 1SG had been surprised that I had come back. He had already reported that I was en route to Germany. I told him there were more important things going on, and I wasn't about to skip out on the rest of the deployment. He then told me that Doc McCollough needed to check me out and to report to him afterwards. He was surprised to see me and we chatted for a bit, before he checked my wound and said I needed to go to Charlie Med every morning to clean it. He was almost surprised they hadn't sent me back with medical orders or profile, until I told him how I had left. He called me crazy and we walked back to the Chus, as he said 1SG would be there.

Upon arriving the incoming alarm rang, and people began running in every direction. However, this time was different. I began to smell something I hadn't sensed in a while. I could almost taste the water of an indoor swimming pool, and he looked at me and said it had to be a chlorine bomb. We looked out and saw the smoke rising from the middle of the base and noticed everyone donning a promask. We ducked inside my Chu and sat down to smoke a cigarette. I turned on an AC unit and we waited out the attack, which was over quickly. There was also a VBIED at the gate, and 1SG ran to my Chu and saw us sitting on my bed smoking. He yelled briefly but looked at Doc and said he needed him. They ran off, and Doc said to wait until he was back. I instinctively looked for my weapon, only to notice it missing, and remembered I hadn't had a rifle since the attack. I got dressed in my new uniform and went to the TOC to meet back up. 1SG asked what profile I was on, and he told him I was a dead man until Charlie Med evaluated me. He told me to relax for the day, and I asked him where I could get a weapon. He had me sign out an HHC rifle and told me to use that until my platoon got back in a few days from the COP. I grabbed it and said I needed to go to the PX, and I was told to go get PVT Fox and bring him with me.

I confusedly looked at Doc and asked what Fox was doing back here. He said he had been injured right after our ambush. Apparently an IED went off when they responded to our attack, and even though he reported being fine, when he got back to the COP he couldn't see out of one eye, and his left hand was numb. They took him to FOB Gabe but couldn't find anything wrong with him. I told Doc it was CPT Roach, so who knew if he just said that because he was an asshole, and in his mind no one was injured. Doc then said he had gone to Charlie Med as well and they had reported him as fine. I shrugged and walked outside, and there was PVT Fox pulling up in a HUMVEE, and as he pulled in, I motioned for him to stay where he was. He was fiddling with the window, but I walked to the passenger seat and told him to just head to the PX. He began protesting that he'd just come back from there and didn't need anything else. I replied that I needed a battle buddy, just got back from Anaconda, and 1SG had said for him to accompany me. He then said he couldn't drive because his eye was acting up, and I told him to trade places.

We left to the PX, and I could see the remains of a few of the chlorine bombs. I resupplied myself, and we went back to the Chu. He began to tell me about his ordeal, and I asked if they had done any scans or other medical tests, and he said they just shined a light in his eye and said it was reacting, but he couldn't see. They had some schedule to go to Anaconda in a few days to check him out. I told him to stay tough and he would be fine. What I didn't understand was that if he was blind and had a gimp hand, why was he by himself and driving all over the FOB?

Fox left back to his Chu, and I found the 1SG in front of his own room and asked what he needed me to do. He told me to relax, Doc said it was a dead man profile, so don't do anything. I told him I was fine, and he said after Charlie Med in the morning, if I was feeling up to it, I could help Stelly put on up armor for some vehicles. I asked if he was fucked up too, and 1SG said no, he just couldn't hack it out there, and they kept him back for details. I told him I would see him after sick call and went back to my room to rack out.

When I woke up, Doc McCollough was knocking at my door and telling me it was time to go. The sun hadn't risen yet, and I could hear familiar sounds of running cadences being sung across the yard. I looked at Doc and asked what the fuck people were doing PT for, and he replied that this place was fucking nuts. We jumped in the vehicle, and I could see units running in formation down the road. Doc started blaring the horn and was using the speaker to tell people this was a medical vehicle and to move out of the way. We showed up, and they took me in and realized that my wound needed cleaning and packing. They asked for my profile and if I had been given anything to pack the wound with and I told them I didn't have anything. The nurse nodded as if she understood the incompetence of the system and told me to lie out on a stretcher.

They unwrapped my wound and had the PA come assess me. He said the front just had to wait for the stitches and staples to fall out, but that the back had opened up and they would need to pack and clean it. The nurse poured Betadine on the wound, causing it to burn, and began using her tools to scrape and clean it.

Just at that moment a man busted through the door, being carried under each arm, screaming in pain and limping like he was severely injured. The two soldiers put him on a cot next to me and walked off. The team came to assess him and asked if he had been injured by a blast or gunfire, and he replied no. He had rolled his ankle while running. They immediately left and told him they would be back in a minute. He seemed surprised and began wailing about the pain. Another nurse told him to relax, someone would be with him in a minute. As they got him a wheelchair and rolled him into the X-ray machine, the nurse who was working on my leg asked how I was holding up. I told her it stung, but to keep going, and just as quickly the man came back around the corner and stood up and sat on the cot. I looked over at him, and he was angry, arms folded, looking around impatiently. It didn't take long for our eyes to lock, and he condescendingly asked what I was doing here during PT hours. I used my thumb to point back at my leg, and the nurse looked at him like he was a child and a fool. He looked away, then asked what was wrong with my leg. I told him I had been shot in an ambush outside the wire and was getting my wound cleaned. He exclaimed that I should be going home, and I told him it wasn't that bad. I was hoping to heal up quick so I could get back out there and kill more terrorists. He fearfully looked the other way and shut up until they came out with his X-ray telling him nothing was broken, he just needed to wrap his ankle. He almost began protesting, until he looked at me again and I smiled back. He then walked out just as Doc walked back in.

I hopped up and left, Doc grabbed my profile and meds, and we went to breakfast. He then took me back to the Chus and told me I was on a dead man's profile, and if he saw me doing anything he would personally kill me. It didn't matter because at that time 1SG was already arguing with SSG Lopez about not having anyone to pull tower guard. I volunteered, saying I was just going to sit around and do nothing, so it shouldn't be that bad. They agreed and sent me off to the towers. I reported in and showed them my profile up front. I wanted to make sure they knew I wasn't supposed to do anything but the tower guard, as they might try and stick me on other details. Then some fat POG sergeant told me I was clearing my SAW wrong, and I blew him off because the attitude of tower guard was ridiculous. They formed everyone up and made us conduct a drill and ceremony maneuver called inspection arms. It was as if none of them had ever conducted PCCs and PCIs. They expected me to do an "inspection arms", a DNC (drill and ceremony) order that you conduct in a parade environment, with an M249, and it was stupid as shit. Think about how ridiculous this is. When in a parade or drill and ceremony environment, you are using rifles (well-polished and unused) in a way to show off and to move around in a ceremony. No one who uses a rifle for a living or goes outside the wire does this. We lock our bolts to the rear, and an NCO clears our weapon, and then you walk by a clearing barrel and dry fire your weapon to prove there are no rounds in it. We were in a war zone, and the POG unit running tower guard was making us act like we were in a ceremony and "checking our weapons". You also don't drill with a machine gun. It was a stupid dog and pony show, and showed just how clueless the people living on the FOB were about the war raging outside.

Next I was assigned a Ugandan soldier as my "battle" for the shift, and the first hour went by much more quickly than with the Georgians, because the Ugandans could speak English. He told me stories of his country and said I should visit. I told him stories of where I was from and said the same. We laughed at jokes and covered for each other on the radio. Then about halfway through the shift, I noticed movement in the field outside our tower. There was a man walking from left to right, and he had something on his back. At first I thought it was a rifle, and then I used the scope to see clearly. He was carrying an RPG and walking in a zigzag pattern across the field. I made sure the 50 was ready to go and sent up an alert to the tower head that I was going to engage one enemy MAM at 200 meters. Right before I pressed the butterfly, a crackle came through and I was told to stop. Apparently QRF had to verify every engagement, and they were sent out to try and find this guy. As they chased him through the field, he would turn and outmaneuver them. We kept giving directions over the ICOM, until finally the QRF asked what tower we even were stationed at. The Ugandan blew raspberries and began what sounded like curse words. He went down the ladder, and I began trying to repeat my location to QRF, until I smelled smoke. I looked down, and the Ugandan had set the brush below the tower on fire to send a smoke signal. I frantically reported this up, to which the tower began frantically questioning what was happening. I told them I was leaving the tower, and they ordered me to stay put. I told them to fuck off, that there were five thousand rounds of HE ammo in that tower and I wasn't about to go up in flames with it. I threw my radio at the Ugandan, grabbed my M4, and walked away as the fire grew to a full blaze in the brush below and enveloped the damn tower and set off the M2, flares, grenades, and everything else. The SOG began screaming at me, and the Ugandan's leadership was actively hitting him. I told the guy to go fuck himself, I was done, and I walked back to the Chus to see my 1SG standing outside with his arms crossed like an angry mother. He was about to lay into me, and I yelled out that I didn't even want to fucking hear it. He asked what had happened, I relayed the events, and he told me I would not be going back to tower guard.

That evening 1SG began to ask me more personal questions and inquire as to why I hadn't left to go back to Germany. I tried telling him some political patriotic shit, and he said he knew better than that. I then told him not only how I felt, but that a big revelation had come from the call I was allowed to make back to the States, to my family on a satellite phone after I came out of surgery. When my mother answered the phone, I told her I had been shot, and right now everything was fine. What I hadn't been expecting was the callousness I would get back in response. It was a very bored, like, "Oh, as a matter of fact, I'm glad you're okay." She even took the time to tell me that since this was an international satellite call, she didn't want to be charged too much and to hurry up my call. I said goodbye and realized my life back home was nonexistent. My parents had always preached self-reliance and had expressed their concern about me joining the Army, but now I was like a black sheep. I couldn't quite pinpoint it at the time, but it felt like I had been cut off and abandoned. I decided that the only real family I had was on this deployment, and I didn't plan on making it home. When 1SG pushed saying I shouldn't want to die, I pushed back, saying it was the only way I got through the day. If I put meaning to my life and have some sort of hope that I might survive, I said, then I'm going to be disappointed and hold back from making tough decisions. If I kept the mindset that every day was my last day on earth and I made it through, then every day I would be surprised. I could stay optimistic, but if I held onto hope that I might live, then there was something holding me back. I was a dead man walking, and I couldn't let myself get out of that mindset. I also brought up the fact that I wasn't going on mid-tour leave, so I needed to be ready for the long haul. He was startled and asked who had told me I wasn't going on mid-tour leave. I reminded him that he was the one who had dished that out as punishment after my Article 15. He told me he was just angry and didn't mean that. I was going home on leave, and to keep my head up. From then on he just needed me to keep track of Fox and Stelly while he was away.

Back on the FOB, still waiting to go forward with my squad after my injury, I was on TOC guard and manning the radio. SSG Lopez had to leave to use the bathroom or get something to eat. I took over just as shit was hitting the fan out in Tahrir. I ended up receiving a report that our tank had taken a catastrophic hit and we had injured soldiers. I had been a part of the Blood Angel platoon and was deeply attached to what was going on. My heart sank into my stomach as they sent up the blotter report that SPC Trussel had been killed in the explosion. I was getting a MEDEVAC report, and everything began to stand still. Just then the sergeant walked back in and jumped on the radio. I walked away, only to see PFC Fox sleeping like a baby on the cot. He had to sleep on it because he said he was suicidal after being examined by a doctor and found to be okay following his exam at Anaconda. It was clear he was completely bullshitting about his event just to get out of the deployment, and it sickened me. I looked down at him with white hot rage, flipped his cot, and began beating the shit out of him. I kicked him, punched him, and screamed how could he stay asleep when one of our brothers just died and shit was hitting the fan. I hated myself even more for being so broke I could not go forward and help everyone out. I fell back and sat on the floor and kicked the cot. PVT Fox cried out to SSG Lopez, and he said he didn't see shit. He told me to go outside and smoke a cigarette and chill out. The next thing I knew, I was back in front of the gym with an M4 preparing for a twenty-one-gun salute.

The platoon was back, and everyone was excited to see me. I spent most of my time reassuring people that I had not in fact gone to Germany. SSG Fili said I was retarded, and I agreed, but that I wasn't ready to go yet. There was a buzz about the company from some news that needed to be put out. We were all weary from the event with Trussel, and from the looks of things, the news didn't seem to be good. CPT Austin confirmed that the news was shit. We were now being held an extra three months in country. Our tour was being extended to fifteen months, and those who had been on ADVON would be spared the detail of REARVON as that would extend our deployment. People began groaning, and some even cried. Stunningly, no one who was on STOPLOSS had been injured or killed. (STOPLOSS is a status when your active time is about to be over, and you would ETS, but the military stops this because of wartime and extends your contract. Remember, everyone signs up for eight years total.) Nearly half of the company were disgruntled hellions who didn't know if they would have made it a few more months, let alone until December. We screamed BOHICA and went our separate ways. I linked up with SSG Haas, and he again remarked that he couldn't believe I would choose to stay in this shit. Davis made a joke about how I brought the bad luck and that's why we were getting extended, and Webb joined in. I told them to fuck off, and we laughed and proceeded to the chow hall and PX.

OUT ON THE TOWN, LATE LAST NIGHT
ME AND SUPERMAN GOT IN A FIGHT
I HIT HIM WITH MY LEFT, I HIT HIM WITH MY RIGHT
I HIT HIM IN THE HEAD WITH SOME KRYPTONITE

I HIT HIM SO HARD, I BASHED IN HIS BRAINS

NOW I'M FUCKING LOIS LANE

CHAPTER 22

Why Go to Disneyland, I've Already Been

I woke up at DFW Airport, waiting for my grandmother to pick me up. I had come home for my fifteen days off and was excited to hit the ground running, despite the long flight. For the first time since October, I was wearing civilian clothes. I was now ready to see my sister and grandmother, then head to Houston. I didn't want to go back there, as I was still confused at my mother's reaction to my injury, but I needed to stop near Fort Hood and pick up my car. Before leaving on deployment, I had asked if my parents would keep my car, and since I didn't want to pay insurance on a car that was just sitting, they wouldn't take it. I then had no other choice but to leave it with Berg. He offered to keep the car at his house until we got back, and I conceded.

However, Berg was now dead, and his wife had made a ten-minute appearance at his funeral, grabbed his SGLI, and run away. What she didn't do was inform the landlord that the car was mine, and he had it impounded. When I called the yard and explained, they said to not worry about the issue, and if I just paid the $20 tow fee, I would get my car back. I thanked them and decided to spend the first half of my leave in Dallas, then go to Houston for the last six days. I wanted to go to Australia, but there was so much bullshit surrounding my reasoning that I just dropped it. I didn't have a plan, or hotel, or passport, and then they asked if I planned to desert. I looked at SSG Haas very angrily at that insinuation and asked if he really thought I was going to desert the Army after coming back from being shot. He said he had seen stupider things in the Army. I blew it off and settled on Dallas. Upon arriving, though, there was an issue. My grandmother was not able to get off work in time, and I was going to have to hang around Dallas for a bit until she arrived. I took a cab downtown and made my way eventually to an Irish pub. I wasn't old enough to drink, but I was hoping someone would feel sorry for me and buy me a drink. The place was packed. I walked to the counter to ask to use the phone. I called my grandmother's work number and told her I would be in downtown Dallas. She told me to have a good time, and I gave the receiver back to the bartender.

Just then the man next to me turned and asked where I was headed, and I told him I had just come in from Iraq and had to wait for my grandmother to pick me up. He was a Vietnam veteran and worked the cameras at CBS. He told me the CEO was having a retirement party and that I needed to meet him. He looked at me and told me that anything on the back of the bar was at my disposal. I could drink for free if he was there. So I drank with the crew and raised glasses to people in Vietnam I didn't even know. I was just glad I could drink. When my grandmother got off work she called the bar, and the CEO told me to not worry about her picking me up. We drank all night until the bar was closing, and I stepped out to see a black limousine. I got in the car and rode to my grandmother's apartment while still drinking in the car. I thanked the CEO for his hospitality and walked to my grandmother's door. I instantly became embarrassed at how inebriated I was. Would she be mad? I had really messed up and started to worry. The door opened, and she hugged me longer and harder than anyone ever had in my life. I sat down on the couch and passed out.

The next day Nance got a hold of me and told me we could hang out. He had been dealing with the fallout of his legal battle, but for the moment he was still in the Army and not in jail. We hung out and got drunk and drove four-wheelers down by a creek. At one point he offered me weed. I had never smoked it in my life.

I smoked a blunt and my mind stopped. I didn't have a single thought in my head for hours. I didn't speak, or joke, or laugh, or anything. It was just like the anesthesia, and I was at peace. Normally my mind races over a hundred miles an hour, allowing me to multitask and hyperfocus on something long enough to fix a problem, then move on to the next one. It's what makes me a storyteller and the bard of the group. I'm always thinking and in my own head. Yet this peaceful bliss allowed me to stop and just observe everything. Nance laughed at me and remarked how I hadn't spoken in hours. I found out that he had attended every funeral of our guys who had died, and some gossip from the rear, but other than that it had been uneventful. I told him I just wanted to get laid and go to a club, but he wouldn't be caught dead at any of that due to his legal problems. He knew that was bad juju and had to stay away. I also found out my sister had moved to New York. Her boyfriend who got her pregnant had left her and taken the child because she was a hot mess. She didn't give a forwarding address or phone and just dropped off the map.

I had my grandmother drop me off at the impound lot and bade her farewell. It was bittersweet as she cried and kissed me. She knew I might not come back the second time and told me to be safe. I told some slight joke and assured her I would be fine. I got my car and drove out to the road to go to Houston. I threw in my CD of Judas Priest and pulled out to Highway 6 ready to leave, and traffic was flying each way at breakneck speed. A woman stopped and waved me by to use the turn lane, I pulled out slowly, and then everything went black. It sounded like a small explosion went off, and my airbag was in my face. I looked to my left and I was further down the road, and to the right was a blue Chevy with a smoking engine. I unclipped my seatbelt and tried to open the door, and it fell on the ground. The truck had struck me from the side and was using the turn lane to try and get around the woman who had stopped for me. I stumbled out of the car and went over to the man in the truck. I tried to make sure he was okay, but he was holding his arm and started screaming. I walked back to my car and saw that the front of the car was crushed and shattered. There sat my Lancer OZ Rally Sports car, and it was completely totaled. I cried out in anger and kicked my tire, and the windshield cracked a little more. At that point someone had come up to me and said they had called 911 and let me use their phone to call my insurance company. Luckily I had reinstated my insurance before I picked up the car. They sent out a tow truck and the cop took my statement, let me use his phone to call my dad, and took me to a motel. I told my dad the address, and he was angry he had to come pick me up near Fort Hood. I apologized, and he even began suggesting I take a cab to Houston. When I told him I couldn't afford a $400 cab ride, he relented and drove up to get me. I went back with little else happening, and as I left my parents' house to go back to Iraq, my mother didn't even look at me when I walked out the door. She had even remarked that I should have gone on vacation somewhere else instead of lying around the house all week. I would like to think it had something to do with her being heartbroken I was leaving back to a war, rather than that she just didn't care because I wasn't following her plan. I stepped on the plane and weirdly felt like I was going home. I smiled, feeling rested and renewed.

It took longer than I wanted to get back to Iraq. I spent a week in Ali Asalim waiting for transport out to Camp Buehring, and another week there before a convoy shipped out to Iraq. I had to ride back with a route clearance team, and it was the absolute worst thing in my life. The black woman who drove the LMTV was an 88M or truck driver, and she was scared to death of everything. She tried to tell me I needed to be on lookout the whole time, and I asked her for what. I didn't have a weapon and didn't see any IEDs. We were in the safest part of Iraq, and they had radioed up that the routes were green that morning. I pulled down my PC over my eyes and tried to rack out, when I heard her screaming. She was going on about some piece of trash and it was an IED. I looked out and it seemed like someone had tossed an old fast food bag out of their truck. I told her such, but she was convinced there was a bomb under the bag. They called EOD out and we waited hours for them to show up. I tried to convince her that I should go kick it, and she said she couldn't let me. She tried telling me if there was a bomb under it, I'd die. I told her if I had to sit in that LMTV for another hour, I'd rather be dead. She didn't find that funny, but I laughed. She told me to pull security while we waited, and I told her I'd use my "finger gun" if any terrorist popped up. She still didn't laugh, and I chuckled and waited. When EOD showed up, they confirmed it was trash and we made our way to Anaconda, stopping for every other piece of trash on the road.

It was probably 0400 the next day when we showed up. I thanked her for the ride and moved to the processing center, where they contacted my unit and gave me a cabin to sleep in. There were ten other guys in there, and they slept all day long. I would go to chow, the gym, the PX, and MWR, but these guys slept 24/7. I would hear the incoming alarm go off and pretend to slow walk to a bunker, and by the time I was there it would quit. The POGs looked like scared rats waiting for the ship to sink underneath the concrete barriers. I laughed at them too and had a total disregard for people who feared any of that.

Soon enough 1SG and SSG Haas were up at Anaconda picking me up, with some other guys, but not before I got a stroke of luck. I knew where the hospital was from my last visit, and they had pilots that would run bombing runs over the country fly out near there. For twenty dollars the pilots would take a flag and fly it over the run and sign the flag. I paid the cash this time around and stowed the flag from the bombing run in my ruck. I linked back up with the company and got ready to move to the FOB. On the way we got a radio message that the whole company needed to move back to the FOB for new orders. We began wondering what was going on, and some joked that we were going to get extended again. We stopped by FOB Gabe on the way and drove behind their med station to pick up water and drinks. The thing I was looking forward to were the Tiger Bombs, as we called them. The actual name was Wild Tiger, and the can had an awesome tiger on the front. A group of military police were carting these energy drinks away, and we began to protest. We yelled that the drinks were for everyone, they couldn't just grab all of them. They replied that they were confiscating them, and we gave them the bird and implied they were lying. However, the 1SG came back and said we had to go.

We pulled up to FOB Warhorse close to dinner chow. I saw the whole company at the Chus and realized this was the first time we had all been together since we had left in October. It was surreal, but we used the time to joke, and tell stories about our parts of our deployment and other things. We gathered around the 1SG and CO's Chu and there was a line of people waiting and my squad joined. The next thing that came down the line was that we were all getting a piss test. SSG Haas looked at me and asked if I was going to be good, since I had just come back from leave. I guess I have the look of someone who would smoke weed, and he told me he would make sure I was good on the test. I told him not to worry, even though I had smoked weed on leave. I was ready to accept whatever punishment if I popped hot on the test, and at this time I just didn't care about much of anything. Yet before we could start the test everything was stopped, and the CO put out official Army orders that placed everyone in our company who was in Iraq at the time on a medical profile to not be tested for drugs. They then told us that the drink we had been guzzling like water, the Tiger Bomb, was now banned, as it had been found to contain cocaine. By this time I had become able to go on missions for forty-eight or seventy-two hours or even longer, and after finishing the drink I would have a burst of energy I had never experienced before. Shortly after you would get angry or irritable for seemingly no real reason. Some guys had hidden them in their barracks, and a full health and welfare inspection was conducted to gather all cans of this liquid cocaine and dispose of it. Davis and I had hidden a good amount at the COP, and once we got back we quickly finished off the pack.

I woke at the COP the next day and walked out to the makeshift showers we had constructed over the last month. As I brushed my teeth the sounds of mortars began in the distance, and I quickly spat and started to move inside. I looked in the yard and saw our company dog, Sandbag, and tried calling him to come with me. He just sat there and looked around a bit before lying down. We called him Sandbag because he refused to move. He was just as apathetic about the mortars as we were. The bombs could land in the yard, and he would stay in one spot sleeping or relaxing. I whistled, but he stayed put, and I turned to walk inside. By now there was a list of actions that occurred during a mortar attack, and the FISTER was already going upstairs to try and ascertain where indirect was coming from. The IA general was going as well and would send his soldiers to check the area with support from Bradleys or tanks. The explosions stopped, and I moved to my daily tasks. I would fill up the generator, check the MRE closet, and take out the trash. After that I would rest before my OP duty began.

We had more people now, and there was time in between tasks, patrols, and guard. I took out the trash, and there at the end of the pile was the biggest enemy I had fought in this war. It was a big rooster, at least two feet tall. He was mean and would fight anyone who came close to the trash pile. We locked eyes, and he began clucking at me. He moved and bobbed his head as he marched forward. Once he was a few feet away from me he charged at full force with his spurs splayed out like a ninja doing a dragon kick. I stepped back and kicked him in his chest with full force, sending him flying against the wall. He came back for a second charge, and I kicked again. By the time he came for a third charge, I grabbed the trash bags and slammed them against his bird body. He decided this was enough and walked away clucking and puffing his chest out like he had won. I flicked him off and threw the gas on the pile and lit it on fire, which caused the rooster to run out of a hole in the wall.

I woke up the next day to find the rooster missing. We had

abused this thing relentlessly over the past few weeks, and Steve had even knocked it out with a shovel at one point. Yet now it was missing, and we couldn't engage in our bit of fun for the day. We looked high and low, and eventually saw the IA guys sitting around eating the rooster. We got mad and started yelling at them and almost got in a fight over it. It's extremely petty and stupid, but when you live like that the smallest things become important, and when they get taken away, you can't help but overreact.

I walked out already pissed and got ready for the next patrol. We had so many people we began volunteering for them, and I would do multiples a day just to get out of the COP. Puffer was on this one with me, and we rounded the corner to see two kids pointing AK-47s right at us. We couldn't think, or even take a second. We raised our rifles up and smoked both. When we moved forward, we found that the AKs were BB guns. I have never been so angry or sad in my life. I broke the guns on the ground and looked around for any adults. They were all gone. These guns weren't real, and the orange cap had been painted over to make it look real. The stocks were of actual wood, and the charging handles even worked. Over the next few weeks reports started going out that this was happening everywhere, and we started taking a second look before firing. The public was outraged, and we had protests outside our gates and at the embassy, but they had only themselves to blame. This wasn't a civil street in some suburban neighborhood. This was Iraq, and this war had been raging since 2003. To let your children run around with guns, whether fake or real, was sentencing them to death, and they knew that. We weren't policing neighborhoods, we were fighting a war, and when someone points a weapon at you, you shoot, or everyone could die.

I understand the police in America say the same thing, and that they have only a split second to decide what's going on, but I can see the statistics and understand that in our streets this only happens to minorities. The police, no matter what they say, also aren't at war with the public. To assume everyone in the public is trying to kill you is what creates these problems. There in Baqubah, in the capitol of Al Qaeda, I was at war with the public, and I did expect anyone to try and kill me at any point. We were invaders, overthrowing their country and occupying their lands. They didn't want us there, any more than they wanted Saddam in power. I cried and felt awful about that event after I came home, but I had to level with the fact that someone put that child in that situation on purpose. Some evil fuck told those kids to hide and then point rifles at soldiers who are actively fighting a war and being shot at daily. Some evil people out there wanted to cause a problem that didn't exist in order to fight us, and it worked.

We found every vendor that was selling fake guns to kids, and we burned them to the ground. The vendors pleaded for mercy, but we held them back at gunpoint. There weren't going to be any more mistakes, and this couldn't go on anymore. We got a call back to go to the FOB. Apparently, some POG unit was about to take over the COP and embassy, and that meant there must be a huge mission at play.

HERE WE GO AGAIN
SAME OLD SHIT AGAIN
MARCHING DOWN THE AVENUE
A FEW MORE DAYS AND WE'LL BE THROUGH
I WON'T HAVE TO LOOK AT YOU
I'LL BE GLAD AND SO WILL YOU

CHAPTER 23

Arrowhead Ripper

The darkness crept in, and fog began to envelope my squad. After the incident, Puffer didn't want to be a team leader anymore and had given it up so I could have the position. I understood. He had arrived late and hadn't experienced most of the bullshit we had gone through. He didn't know the area, and the incident choked him up. We moved through a field and into a palm grove, while I heard different elements on the radio sounding off at various checkpoints. I only had an assault pack and a radio, which was a nice change of weight, and I was trying my best to listen in to the shitty hand mic given to me. All the voices were barely audible, and I was straining to make out what was passing along the net. We made our way through the grove until we came to a town that was in the middle of the thickest part.

This was the reason we were here. This area had been disputed by each battalion as to whose AO it was and had yet to have been cleared by anyone. We were now part of a three-day-long mission to clear this entire town and area, as the suspicion was that Al Qaeda was working here due to our lack of presence. I looked ahead to see that people were stacking outside the building and preparing to enter and clear, and something caught my eye. One guy was using a tac light on his rifle. The door was kicked in, and the team poured through the door. I then was nearly blinded by the light that guy used to shine down the hallway. Before I could come over the net and tell them to shut it off, bullets rang down the hallway and caught the kid in the neck. We quickly sent another squad in the building, and they killed all enemy personnel inside. We stacked around the house and pulled security while a medevac was called in. They decided to send us up forward, as the town we were outside of was being bombed by JDAMs as we approached. It took all night to move to the highway we were supposed to use as our start point, but we finally did. I told my team to remove the Velcro from their KPOD and show their IR squares. Hopefully with the whole company doing this, the plane would see us at night and not drop a bomb on our heads. We were relaxing, and at this point I found Melsha was falling behind and couldn't hold the SAW all night. He was complaining so I told him just forget it, and I traded the SAW with my M4 and assault pack. I also took his ruck and moved the radio and my rounds in the ruck. I had two extra drums of ammo and we switched kits, and I found out he only had 600 rounds. I had told him to carry 1200. I huffed that he couldn't carry the weight, but right now I needed to stay focused. The radio started chatter that the bombs were about to drop, and we needed to stay down. I was leaning against the ditch when I felt the force of wind blow me forward. I did a somersault in the air and landed on my face. People began gathering around to see if I was okay, and Puffer started laughing. He had seen the whole thing from the other side of the ditch and said I had flown like a ninja. My pride was hurt, but we still had another three days to go for this mission, and this is how it was already starting.

After the bombings we began to walk towards the town in the distance. All night the ground under my feet was crunching, and an awful smell emanated from the ash as it lifted into the air with each step. It wasn't until the sun came up that I noticed shapes coming out of the ground and realized what we had been walking over. Distorted hands and feet were extending from the ground, and I could see the outlines of bodies. These were the combatants that had been killed in the attacks. I'll never forget the smell of burning flesh. We walked through another field behind our LT and came upon a donkey blocking the path. It was trying to turn and kick everyone, and we poked it on either side as we walked by to confuse it. It became more irritated and began to haw and huff, and our LT was afraid to walk by. He pulled his pistol out and shot the donkey in the head. It began bleating and kicking, and his fear took over as he unloaded the mag in the creature's face. SSG Haas had enough and put the creature out of its misery. He yelled at the LT, and we moved forward. We could finally see the town. People were scurrying across the roads and buildings. It was time to form up, and we moved into the corner of the town. An eerie silence fell over the town, and older men started coming out of the buildings. Each one had an AK, and we quickly noticed we were outnumbered. Our presence wasn't wanted, and they began trying to encircle us. I don't know where the first shot came from, but bees suddenly filled the air, and we returned fire. They chased us through the streets, and we fell back close to a Bradley that opened up on the fighters. There was no room in the Bradley, and they told us to continue falling back to the COP. We ran and kept running while turning and shooting, until we had bounded all the way back to the COP. At this point the remainder of the mob began firing on the COP, and they opened with everything. Mk 19s rang out next to .50 cals, and nothing was hitting anyone. The POGs at the OP couldn't shoot worth shit, and I dropped the last of my drums of ammunition as my team crossed over into the COP. A Bradley came around the corner and finished off the rest of the fighters.

We linked back up with the platoon and began clearing the streets. We took more fire and moved to rooftops to hold positions and call for fire from helicopters. I found myself clearing with SGT Rojas and CPL McGrath for a moment and watched as he grabbed a Mossberg shotgun and tried to shoot the handle of a metal door. I was about to say something, but I remembered that Rojas and McGrath had broken Webb's jaw in the rear for some petty indiscretion, and now payback was a bitch. As he shot the door with bird shot, the pellets traveled back up the barrel and took his nose clean off. We had to call in a medevac for him, and he was driven back to FOB Gabe. We found out SPC Nash, one of the FISTERs, had been killed in an IED attack as well. We didn't have time to think or do anything but continue. We cleared every house and inch of that town and took out all resistance.

As the day started winding down, we were reporting up that we were black on water. Everyone had finished their last drop, and we needed some resupply to keep going. They wanted us to march back to the COP, and all vehicles were claiming they had to provide security for the command teams. We protested and found a street with good cover to hold up. Just as Webb turned the corner to secure the rear of the formation he knelt and his crotch ripped, exposing his entire private area. It was also at this point that SGM Harris pulled up and began demanding we keep marching, and Webb threw his weapon down. He told him we had been black on water for over twelve hours and weren't marching anywhere until we got a resupply of water. When SGM Harris began yelling, Webb sat down in the middle of the road and told him he wasn't moving until he got water. A medic came out from another vehicle and identified that much of our platoon was about to become heat casualties and needed water if we were to continue. SGM Harris finally gave in, and Webb and a few others were CASEVACed back to the COP and given IVs. I had to give it to him, he didn't give a fuck, and being extended on STOPLOSS wasn't going to help him give a fuck. The rest of us moved on and were walking towards a part of the town based on active intel. There was a Kurdish neighborhood, and they had identified a man who wasn't native to this place. We asked if it was Ali Baba, and they said no, he was different. We came by a set of buildings until it struck me. I was looking at a Honda Civic. Cars in Iraq were mostly what we called Bongo trucks, the little blue trucks that cross most of Europe and Asia. We saw a few sedans here and there, but I had never seen a Honda. We knocked door by door, and another platoon found the second sniper brother we had been looking for. That night the FOB was on total blackout, and Black Hawk helicopters landed and took this prisoner to some undisclosed location. That was the last we heard about the sniper brothers.

After the full clearing of the village, things died down. We moved through Baqubah on regular patrols and participated in more missions with a civil relief effort. We gave toys to kids and clothes and food to local nationals. We guarded civil projects to build roads and conducted meetings with community leaders. They sat in our COP and spoke to the commander about their grievances, and he offered what aid the unit would allow. At one point we helped the SF (special forces) conduct a recon and intel mission that attempted to bug a house in Baqubah, but it didn't gather any results. This was probably because the SF team that showed up was trying to blend in with our unit and failed miserably. They hadn't worn traditional uniforms for most of their deployment and spoke candidly about how shitty our gear was. When patrolling through the town, they were visible because they didn't fall in formation with us, and we didn't stop at any other house than the one they wanted to bug. The entire town could see that something was up, and after we overstayed our welcome at the house, we were rushed out by the people that lived there. The next day when we returned the bugs had been found and removed, and the people the SF team wanted to interview were gone.

However, the previous night I had been able to shoot the Barret .50 cal sniper rifle. A local had moved past the barriers we set out on the OP to guard a blind spot for our vehicles and cut the concertina wire while moving through. I couldn't tell if the person had a weapon or not and called it up on the radio. He continued to crawl over the second of three barriers. Both were highly illuminated, with warning signs to stop or we would shoot. There was a Bradley in the area, and they came over and began blaring on the speaker that the man should turn back around. He maintained his silence and continued moving over the second barrier to the third concertina wire. At this point one of the SF guys had woken up and nearly made me piss myself asking if I wanted to shoot the Barret. He had been sleeping in a dark corner of the OP so I couldn't see him, and

I sheepishly replied asking if a bear shits in the woods. He laughed and set it up for me and had me get in the correct firing position. He called up on the radio so I could maintain my sight picture with his IR scope, and then they confirmed to fire if the man cut through the third concertina wire. He broke through, and I let one .50 cal bullet fly through the night sky. The shot reverberated against the walls, and my head began vibrating like a bell had been rung. The man's leg went flying in the air, and he lay on the ground screaming in pain. I let another round fly and he went fully limp. I cleared the rifle and reported up the BDA and lit a celebratory cigarette.

The next major mission we were stuck on was protecting elementary school girls from violence. Of all the patrols and missions this one stuck out the most, because of the way many of the fathers looked at their children. Kids hadn't had the ability to go to school in almost a decade. Much of the population couldn't read or write, and it showed in the level of cognitive ability of most adults we interacted with. The part that was abhorrent to these families was that girls were going to school. Each day it felt like being at the doors of the school the Little Rock Nine or the black children from Mansfield, Texas had tried to attend. These men of the community stood outside and stared and howled at these girls on the first day, held signs condemning them to death, spat in their direction. I couldn't stand idly by. I thought about the white people that fought against school integration for black children. I thought that if I had been in the same situation, I would have been better. The first day a few of the fathers attempted to break our line to get at the girls, and we butt stroked them in the face and put them on the ground. We kicked them, pushed them back out bloody and with broken bones, gave some of them to the IA for interrogation. We knew the general would conduct one of his special sessions on these men, and they wouldn't be a problem afterwards. I spat in the face of one and laughed as the man tearfully fell to the ground. If they got close to the line we struck them with our weapons, and if they touched us, we pulled them behind the line, where IA soldiers viciously beat them. This had the effect of showing them their own government and people were not going to put up with their zealous bullshit.

The next day, most of the crowd had calmed down and there was only one incident. I had seen a man eyeing a girl the previous day. She was at least seven years old and had a little Dora the Explorer backpack that she wore as she happily sang. This man looked at this girl with hate in his eyes. He spat in her direction and kept following in the crowd as she made her way to the classroom. He returned in the afternoon to mean mug her again, and he took her hand at the end of the line and walked off. I grabbed his hand and pushed him back, until the interpreter confirmed that this was his daughter. He looked at me and smiled, because now he could inflict his hate on his own daughter, and no one could stop him. I looked at him with rage, and in my mind was daring him to do something. The next day the little girl wouldn't look up, and I stopped her and looked at her face. She had a huge black eye, and the man looked at me laughing. He spat on the ground and looked as though he had won. He thought if he didn't mess with our line he was untouchable, so I broke rank as SSG Haas yelled at me. I walked up to the man, grabbed him by the ear, and brought him behind us, and the IA began to kick the shit out of him. As they held him down, I kicked the man repeatedly in the privates and hit and spat in his face until it bled. We then pulled him out in front of the line and paraded him up and down in front of the other fathers. We had Snake call out asking if any of them wanted to look like this man. Was it worth denying their daughters the right to read? If so, then the next man should stand up. They all went home, and from then on our security detail was uneventful.

The final two major events of the deployment occurred in September 2007. The Nasir Square incident was a huge indictment of all soldiers and mercenaries in Iraq at the time. Though people were shocked and rightfully disturbed by it, any soldier deployed could have told you it was going to happen. We were en route to Anaconda at the time and decided to divert after hearing about it.

We certainly didn't want to get caught up trying to protect mercenaries who didn't give a shit about us and caused this issue in the first place. We headed back to our FOB and spent the night there, just in case we got called up for some FORSCOM mission. My thought was that they deserved it, because these people were just abusing the war and our tax dollars for their own enrichment. Every day I saw more and more contractors showing up to take over work from service members. By the end of the deployment the only thing POGs seemed to be doing was tower guard, and the bases we were building and improving were being constructed by locals under the management of civilian corporations. The food we ate was made and delivered by Halliburton and KBR, and our expensive anti-IED equipment was contracted out as well. At one point I watched as a contractor used a flat roller to crush thousands of "defective" ACOGs so they could report them as destroyed and have the Army issue new ones. I asked if I could have one, and the contractor told me with a straight face that I would be engaging in fraud if I took one of these ACOGs for personal use. Every one of those sites could have been rebuilt and fixed, but they needed their money. So, forgive me if I don't feel bad for a few mercenaries rotting in jail because they got caught doing what they normally did.

On the way back to the FOB we had our own incident. We were going through checkpoints and almost at the FOB, when an MRAP came flying around the corner and struck our Bradley. I was in the HUMVEE behind the Bradley, and it took me by surprise. I instinctively turned my turret around and pointed at the vehicle, only to watch it roll and flip twice. It lay on its side, and IA began to scramble and exit the vehicle. I could hear screaming over the radio, and SSG Haas said the IA had blown through the checkpoint. Normally you slowed down and yielded if the guard at the checkpoint stopped you, but with our entire convoy being the larger traffic he pointed for us to go through. The IA ran around the MRAP like it was a clown car and, while moving to the gunner hatch, I saw the man fall out completely limp. He had died on impact and wasn't wearing a harness or belt. His body slumped like silly putty as they moved it out of the hatch, and they were screaming and crying. One IA member pointed his AK at us, and I drew down with my .50 cal and screamed at him. He dropped his weapon and it discharged. He jumped on the ground, and we just rolled our eyes. This group of idiots was going to supposedly take over the country and protect the nation. We drove off, reported the incident, and arrived at the FOB.

As we drove to the motor pool, I could see a few of the mechanics huddled around something. Davis and I jumped out to see a bunch of jackass mechanics playing with a rusted AK-47, pointing it at one another. They looked just like the idiot IA members, and I couldn't let it go. I started screaming at them. They all turned around, and I could see one was an E-5. I told them to stop fucking around like that, not to play with a weapon like it's a toy. One guy piped up saying the thing was rusted and couldn't fire. They found it buried in the silt, and it didn't even have its buttstock anymore. In one of my few moments of awesomeness, I yanked the rifle out of the kid's hand, racked the charging handle, and shot a burst of rounds into the dirt. I remarked that these rifles were meant to survive everything and repeated that it wasn't a toy. I handed it back to the kid, told him to clear it, and said if I caught him pointing it at anyone I'd box his ears. I didn't have time for any bullshit, and I was over this unit and the idiocy we had been abused with over the last year and a half. I gave the NCO a look of disappointment, and he quickly grabbed the AK. We fixed up the vehicle and began getting the local PNN from other platoons and mechanics. They laughed because they had seen what I did and told me I was fucking wild. We found out that the motor pool sergeant was going to get some Bronze Star with valor, even though the only mission he could have left the FOB for he had refused to do so, and that every NCO in the battalion was going to receive a Bronze Star. I remarked about how I didn't give a shit and hadn't done this for some stupid Boy Scout badge, and I meant it.

The deployment was a shit show, we didn't solve any problems, and we didn't stop terrorism. What the fuck were we patting ourselves on the back for? Every turn of events happened because we were undermanned, poorly planned, and led by barely competent people who if they were smart enough were only keeping things together by a thread. I wasn't part of big meetings, or mission planning, or major coordination of anything, and I could see how fucked up things were. Of course, I myself was fucked up. I had finally received my SPC the last four months of the deployment. I couldn't get promoted and maintain rank to save my life. I acted out, was a fool, violated regulations for a laugh. How was anyone to expect anything less from this organization, when I was the average joe that existed there? I could see and understand that we had only been sent overseas to kill a bunch of brown people who couldn't speak our language so that Halliburton and KBR could have their stock rise by half a percent. We didn't respect these people, and they didn't seem intelligent; their ways were backwards. They ate with their hands, hated dogs, killed women who had been raped, allowed grown men to have boy lovers but stoned gay men to death.

You can't see the humanity in their eyes and can't understand why they do the things they do. But you can't lie to yourself and pretend you think those kinds of people set up the 9/11 attacks. You can't lie to yourself and say these people are constantly plotting to overthrow your country and ruin your way of life. What does it say about you? That you're over here firing a 25mm cannon at some random guy who's just pissed because you're in his country. You're trying to fight for freedom? Whose freedom are you doing this for, and if it's theirs why don't they care? If it's yours, then how come there aren't more attacks in America? What does it say that we stopped giving a shit, and just liked it?

OH HAIL OH HAIL INFANTRY
QUEEN OF BATTLE FOLLOW ME!
AN AIRBORNE RANGER LIFE FOR ME
OH, NOTHING IN THIS LIFE IS FREE
OH HAIL OH HAIL INFANTRY

E4 MAFIA 4 LIFE

CHAPTER 24

I Thought I Was Already Home

I stood in line in for the greatest McDonald's that has ever existed, and that's saying a lot because I've been to the McDonald's in Austria that overlooks the most scenic mountains and sells Italian coffee and beer. I was ready to overpay for two Big Macs and a supersized order of fries. I was on REARVON even though we had been promised if we were on ADVON we wouldn't be on REARVON. This put my deployment time to seventeen months, and it didn't even reflect on my DD 214 when I left. That's a stab at anyone who claims they've been "deployed" to Kuwait. The Army didn't even think my two extra months counted as a deployment, so I certainly don't think anyone who rode out the war in Kuwait had a proper deployment either.

The downtime before returning to Fort Hood left too much time for internal thoughts and personal reflection. The air stunk with anxiety that was worse than when we first stepped foot outside the wire. It was almost like the real deployment was about to start. I had been told prior that I would end up with PTSD and become a nonfunctioning member of society, and the cases of people wanting to seek behavioral health were rising in the unit. I began to ponder if anyone else had the feeling that they didn't deserve to live. I didn't have any feelings at that time that made me morally question my role in the war on terror, but I did wonder why I was alive, and felt that anyone who meant a damn had died. It's the shitbags like me who continue to live, and the adage that the good die young rang through my head. I thought of all the people who had died or been permanently injured overseas and how they didn't deserve it. Most of them were considered decent soldiers, yet I had witnessed the best soldiers fall out, desert the Army, and die or get injured. The guys who can't keep up, the ones who can't get their shit together, the guys who maybe weigh a hundred pounds soaking wet, they all made it. The Stellys, Malitzes, and Fullers of the world make it across the finish line and raise up their gold medal and crack a bottle of champagne. So what does that say about you, if you're right there with them in the end? I remember walking past a trailer that had a Special Forces recruiting center in it, and began to turn in the door. There were signs on the rails that said if you signed up to extend your deployment you would be able to join SF. SSG Haas caught me and told me to walk away. He said this life wasn't for me, and I needed to go back to the States. I told him I hadn't had enough of the deployment and didn't want to go back stateside. I couldn't stand the regular army, and I couldn't stand garrison. The thin veil of society was enforced there, and I didn't want any part of it anymore. I've never let go of that feeling, and since then have always felt out of place. In Iraq I was free, and I was worth something. The idea that I needed to go back and be a good boy, and pay taxes, and live a normal life, was daunting. What kind of world was I going back to fit into and be accepted in anymore?

But I didn't go in the trailer, and I followed the unit back to the States. We boarded the plane and arrived in Edinburgh before the sun came up. We were allowed to go out with the crew workers and smoke cigarettes and laugh at their funny Scottish accents, and they handed us a few beers. We saw SSG Haas, and he told us just to not get drunk, we didn't need an international incident before we got back stateside. Once on the ground we came out to a huge pavilion where our families were waiting on us. They cheered, and as we stood at attention a celebratory cannonball was fired on the parade field. This was the last straw for one of our guys. He had worn his KPOD every day in Iraq and was convinced that if he took it off, something bad would happen. I could see his perspective, because he had been in some serious shit. He had been mortared multiple times, and at the end of our four months' stay at the COP he took his helmet off to shower and the showers were shot up by an NCO who cleared the main gun of the Bradley incorrectly. He was the only one in there, and it happened right as he took off his KPOD. He kept that helmet on the entire way home, and even on the plane ride back. When he was on the parade field, he finally removed his dome piece and sighed with relief, until that damn cannon shot off. He put it back on and ran down the road. It took a team of MPs to get him and move him to the fourth floor of the medical center, the psych ward. I couldn't help but laugh at the chaos and feel relieved I wasn't as messed up.

I walked around the field until my best friend from home popped up out of nowhere. James Wilkie had ridden up with my parents and was there to bring me back home. I was really surprised that they even knew when I was going to be back, and they said they had gotten a call from the unit FRG (Family Readiness Group) telling them our arrival time. I started to feel a bit better, until I remembered the life lesson I lived by. Don't ever get too happy and always expect the worst thing to happen, that way you'll never be disappointed.

We were formed back up and told we weren't allowed to leave. We had to go to an immediate reverse SRP (Soldier Readiness Processing) and were going to be held until the next weekend. We had leave forms to sign, passes to fill out, and medical examinations that needed to occur. They always forget to show you this bullshit in the movies, and wouldn't that be such a great ending to *Top Gun* or *American Sniper*? Watch as the hero comes back and sees his family that came from miles away to cheer his return home, only to be shuffled to reverse SRP and have to tell them he'll be home next weekend! You must go through the grind and get processed before you can just up and leave anywhere, and my parents were in full protest about the event. They didn't realize the Army doesn't give a fuck, and no matter how they kept commenting on how they wouldn't even have come if they had known they couldn't take me back home that weekend, it wouldn't change anything. They then said they wouldn't be back next week to take me home, and I would have to find another way. Before I could even begin to barter, James said he'd come get me, and I gave him a hug and thanked him. We all formed up and were brought into a main hall so we could be issued room keys and have a quick reentry back to the unit in the States. I began to laugh. People around me were crying and raging out. I heard all sorts of protests, but I just walked to the tent and went through the grinder. I had gotten caught off guard and let my happiness out for a moment, and the Army was there to remind me that I wasn't allowed to have that.

I DON'T KNOW WHAT I DID WRONG
AND IT WON'T BE LONG
TILL I TILL I
TILL I GET ON BACK HOME

CHAPTER 25

I'll Never Go Back Home

The next weekend James Wilkie arrived to pick me up from Fort Hood. I hadn't gotten a car yet, and I was sitting on a shit ton of deployment cash. Unless you deploy while you're married, you come back with an amazing nest egg of cash you didn't get to spend while in Iraq. There were only so many packs of smokes available at the PX, and you could only pull out so much cash from the FOB finance department. I think I pulled the maximum amount when I first hit Iraq and didn't pull any more money out for at least six months. I also sent my deployment money to a savings account and didn't touch it on leave. My first weekend I bought a brand-new PlayStation and one of those new iPhones that had just hit the market. The week went by, and James came and picked me up, and I didn't have much to say at the time. I couldn't figure out what exactly was weird, but I was lucky Webb had decided to tag along with us, so he spiced up the silence with goofy conversation. There were a few people that took passes to go home for just a few days, and I couldn't understand that either. We would be getting thirty days leave, so why waste the tickets now? It didn't matter. We went back to Houston for the weekend, and we drank and partied until the sun came up.

When we came back to Fort Hood on Sunday, SSG Haas was freaking out and glad to see us. He couldn't get hold of us over the weekend and told us that this weekend Fort Hood had basically destroyed itself. There's always rabblerousing when a unit gets back from deployment because they've been pent up for so long it just spills over in the end. When we had our Monday morning formation, SGM Harris just sat on the ground and held his head in his hands and exclaimed that more people had died that weekend than on the deployment.

We had a running joke about Fort Hood that involved the sign on the outside. You can see a counter that tells you how many days since the last motor vehicle death occurred in a unit, and while we were in the rear it never got over three days before resetting. It was such an issue that the Post General had issued a seven-day pass to all soldiers on base if we could get over thirty days. We never collected on the offer, but when we came back from deployment the counter was up to 360 days without a death. The joke was that it would be zero after the weekend, and we were right. SGM Harris detailed how many motor vehicle deaths had occurred, how many people had died from drug overdoses, a few suicides, and most interestingly a nine-person knife fight outside Starlight Station that resulted in some deaths. The scuttlebutt was that they were all fighting over a stripper. The tale was as old as time itself, and I basked in the absolute hypocrisy of it all. The American hero, Maverick, returns home to find his wife has taken all his money out of his account, served him divorce papers, and got in Jodie's car with their kids and dog. She hates the dog, too. Where is this at the end of those military movies? Where's this in the tale of patriotic fanfare about our brave soldiers? As some soldier sits in a barracks room and puts a gun in his mouth, or sits in jail overnight because he yelled at his wife for lying, cheating, and stealing, where's that part? We see the guy get the medal and some asshole thanks him for his service, but this is nowhere at the end of any movie I've seen. The normal thing that occurs is that you come back to trash, your family has learned to live without you, the people who "care" about you act like you're a parasite in your own house. I saw so many guys who got married to their high school sweetheart before deployment divorced before the year was out. I've had brothers from this deployment rack up six marriages along the way. More people died coming back from combat than died in combat. Remember that.

Things would slow down, the world would keep turning and I would try to move on, but I don't think I ever will really be able to do that. I said at the beginning that I'm not a hero, and no one who deployed with me is either. We didn't storm the beaches or blast through the eastern front to stop some egomaniac who wanted world domination. We didn't stop terrorism or make anyone in the United States freer. Hell, just the fact that you must expect to get molested if you ever want to fly in a plane should tell you how much freedom we have today versus prior to the war in Iraq. If any of this disturbs you, I want you to know that's the point. It's good, and it means you have a bit of humanity left in you. We did horrible things, and fucked shit up, and didn't save the day. What you did was that you expected young men to go and fight a war in a country you probably couldn't identify on a map. You expected eighteen-year-old kids to go fix the cultural problems of a region of the world that can't be fixed, and you pawned all your responsibility and liability off on us. We were set up to fail, and you, the American society of today, haven't brought anyone to justice over it. We still allow war criminals like Dick Cheney to steal their third heart from some poor deserving soul, and George W. Bush gets to paint shitty paintings, and another war hawk, Hillary Clinton, went on to almost become the first woman President of the United States.

The people that end up being heroes are PVT Fox. He went before Congress to fight for better medical care for soldiers and veterans, but he was still lying about all his injuries and did everything he could to leave the deployment. Hell, we had a guy lose an eye and he came back to Iraq, but Fox didn't want the brotherhood, and he didn't want the suck. So his family, who were well connected with George Bush, made some bullshit fanfare in front of Congress, and they gave him a standing ovation. The men who fought and died so they could keep all their corrupt deals, and lies, and bullshit going were tossed out in the cold.

Things only get worse once you leave the service after war. Time stands still for who you were, but everyone else moves on. They live in a world you will never touch again but will always be somewhat desperate to rejoin. It's always those who claim to respect you the most who respect you the least. Your soul was crucified for American hegemony around the globe, but you aren't meant to live in society anymore. People smile, they talk down to you, and they see a young man or a kid. They don't know, or if they do they forget, that you literally spent a year and a half of your life hunting human beings, killing, and watching your family die. I lived a thousand lifetimes in the span of a year, yet veterans get this fake show of respect when we return home. I almost wish I could have been spat on like the veterans of Vietnam, because at least that's honest. There are very few jobs in the Army where you can leave and walk on the civilian version with no extra certification, licensing payments, or strings attached. So remember that when you see 22: the number of vets who commit suicide daily. Or ask yourself why there are homeless veterans. It's because everything we did in service for you during the Army no longer counts as real world experience. We weren't certified or good enough for you.

You hear people say that your experience is something no one can take away from you, but really they're telling you it has no value. My first enlistment was four years in the Army and deployed overseas as an infantryman, so can I get a decent job? Well, I found out that I couldn't be an armed security guard because I wasn't certified, and my first boss at a retail boat store reminded me that I only had a high school education. I would find out when I certified as an EMT that the medics who come out of the Army aren't promoted to paramedic, and up until a few years ago, anyone who drove military vehicles couldn't get a commercial driver's license unless they went through the same shit as everyone else. You have your G.I. Bill for college, and that's great for some people. Others don't want anything to do with higher education, and there are tons of trades and trade schools that don't qualify for your G.I. Bill. I tried to get back with Jennifer Lemus from high school, and she didn't want anything to do with me anymore. She was the only woman I had really loved and had a connection with, and I didn't know where any other pieces of my life existed at the time. I still wonder sometimes if I would have healed or been better if we had gotten back together, as she was the only person who understood me in high school and was the only positive female relationship I ever had. She told me I hadn't changed at all, and I couldn't believe it. I couldn't believe that the most shocking event in our culture hadn't changed me, and yet I was also trying to wear it as a badge of honor. The machine didn't change me or take my soul, and I wouldn't find out until years later how untrue that really was.

I went to the VA after my first deployment and tried to see if I qualified for disability, and literally had a man tell me that I didn't qualify after being shot for my country. He qualified, because he broke his ankle in a road march back in '82 on some unknown base, but the only way I was going to receive any benefits was if I told him I had PTSD and they went through the long process of proving it. I found myself bouncing between odd jobs while casually going to school against my wishes. I just wanted the monthly check and tried my hardest to bury everything with alcohol. The funny part is I was too poor to drink like I had done in the Army. I hadn't gotten the money, the social satisfaction, or the girl. After everything, I was just another shitbag.

Is it really a surprise that twenty-two veterans kill themselves every day, and most combat vets are homeless? The only work you qualify for when you leave is bullshit minimum wage jobs or worse. Can my deployment help me rejoin society at a level above poverty? Can the experiences of war help me make something of myself and take care of a family? Can the blood spilled by my brothers allow me to be socially uplifted to the level of the hero that society says I am? Well, it's an accomplishment no one can take away from you. Now do you want fries with that?

US ARMY
BARTA
U.S.ARMY
DOC
ROSS
SPENCER
SIEGER
US ARMY

About the Author

Stephen Russell, a former airborne infantryman with a decade of military service, brings a unique perspective shaped by his deployment with Bco 1/12 3HBCT 1st Cav DIV to Baqubah, Iraq, during the surge in 2006-2007. Beyond the battlefield, he's a versatile artist, created the comic series *Tales of Nihilism* (globalcomix.com/a/tales-of-nihilism), and is open for traditional art commissions. He believes the U.S. hasn't fully reckoned with the aftermath of 9/11, emphasizing the need to confront our collective anger and hate.

About Blue Ear Books

In 2009 journalist Ethan Casey, author of the landmark travel book *Alive and Well in Pakistan* (praised as "magnificent" by Ahmed Rashid, author of *Taliban* and *Descent into Chaos*), decided to take control of the publication and promotion of his own books, rather than remain subject to the whims and vicissitudes of a perpetually and radically changing publishing industry.

That led to the founding of Blue Ear Books, through which Casey and a small group of colleagues publish a select list of books by authors with distinctive professional and personal perspectives on the international world in the 21st century. To learn more, visit:

www.blueearbooks.com

Made in the USA
Monee, IL
04 November 2026

67377575R00163